# The Global Equivalence Ratio Concept and the Prediction of Carbon Monoxide Formation in Enclosure Fires

William M. Pitts

I0482734

Building and Fire Research Laboratory
National Institute of Standards and Technology
Gaithersburg, MD 20899

June 1994

U.S. Department of Commerce
Ronald H. Brown, *Secretary*
Technology Administration
Mary L. Good, *Under Secretary for Technology*
National Institute of Standards and Technology
Arati Prabhakar, *Director*

# TABLE OF CONTENTS

# TABLE OF CONTENTS

# LIST OF TABLES

v

# LIST OF FIGURES

# LIST OF FIGURES

# LIST OF FIGURES

# LIST OF FIGURES

## LIST OF FIGURES

# LIST OF FIGURES

# LIST OF FIGURES

# LIST OF FIGURES

# ABSTRACT

This report summarizes a large number of investigations designed to characterize the formation of carbon monoxide (CO) in enclosure fires--the most important factor in fire deaths. It includes the first complete review and analysis of the studies which form the basis for the global equivalence ratio (GER) concept. Past and very recent (some as yet unpublished) investigations of CO formation in enclosure fires are reviewed. Based on the findings, two completely new mechanisms for the formation of CO, in addition to the quenching of a fire plume by a rich upper layer which is described by the GER concept, are identified. The first is the result of reaction between rich flame gases and air which is entrained directly into the upper layer of an enclosure fire. Detailed chemical modeling studies have shown that CO will be generated by these reactions. The second is due to the direct generation of CO during the pyrolysis of oxygenated polymers (such as wood) which are located in highly vitiated, high temperature upper layers. The findings of these studies form the basis of an analysis which provides the guidelines for when the use of the GER concept is appropriate for predicting CO formation in enclosure fires. It is concluded that there are limited conditions for which such use is justified. Unfortunately, these conditions do not include the types of fires which are responsible for the majority of fire deaths in building fires.

# I. INTRODUCTION

Roughly two thirds of all deaths resulting from enclosure fires can be attributed to the presence of carbon monoxide (CO) [1],[2] which is known to be the dominant toxicant in fire deaths [3]. The mechanisms responsible for the generation of high concentrations of CO in fires are poorly understood. This work summarizes the results thus far of a long-term program which is seeking to develop an understanding of and predictive capability for the generation of CO in fires. [4]

The majority of the effort has focused on providing answers to two research questions which are central to the goal of the effort. [5] These questions can be summarized as:

1.  Does vitiation of the oxygen supply (normally air) for a diffusion flame lead to the generation of higher concentrations of CO than burning in unvitiated air?

2.  Can the generation behavior of CO observed in hood experiments designed to model two-layer burning be extended to predict CO generation in actual enclosure fires?

Vitiation refers to a reduction in the oxygen concentration as a result of mixing combustion products with the air.

The first question has been addressed in earlier published work. [6],[7],[8]. It has been demonstrated that the generation of CO for fully ventilated (i.e., having sufficient oxygen available for complete combustion of the supplied fuel) fires is only minimally affected by vitiation down to levels sufficient to cause extinction of the combustion itself. Under these conditions the concentrations of CO measured tend to be quite low. On this

2

basis it was concluded that vitiation effects on the generation of CO in enclosure fires would be minimal.

This work focuses on the answer to the second question. In the following section the problem of CO formation in enclosure fires is discussed and the need for accurate prediction methods, which is the ultimate motivation for the research effort, is discussed. In Section III the original studies of simplified two-layer fires which led to the development of what is now called the "Global Equivalence Ratio (GER) Concept" are summarized and the emphasis on question 2 above is justified. Section IV discusses the findings and implications of a study which used a detailed chemical-kinetic model to predict the reaction behavior of combustion gas mixtures observed in the simplified fires. Section V summarizes the findings of investigations of fire species production in reduced-scale enclosures both at NIST and other laboratories. Recent experiments at the Virginia Polytechnic Institute and State University (VPISU) and NIST are emphasized. The flow behavior for the reduced-scale enclosure fires at NIST has been calculated using a field model. New mechanisms for the generation of CO in enclosure fires are identified. In Section VI the results summarized earlier form the basis of an analysis designed to assess the appropriateness of using the GER concept for predicting the generation of CO in enclosure fires. The GER concept appears to be applicable only for very specific types of fires which are not typical of those believed to be most life threatening. Limitations of the current work and research areas requiring further effort are discussed in Section VII. The final section (VIII) is a summary of the conclusions.

## II. CO FORMATION IN ENCLOSURE FIRES

A focus of the Fire Program within BFRL has been and remains the development of a computer-supported methodology known as *Hazard I* [9] which is capable of predicting life-safety hazards associated with enclosure fires. A major limitation of the computation is that the production rate of molecular species as a function of fuel-loss rate must be input as a parameter. The understanding of the mechanisms responsible for the formation of CO in fires is too limited to allow accurate predictions of generation rates for this species. This is particularly true with regard to extremely intense fires where such high generation rates of gaseous fuel occur (either directly as gas, vaporized liquid, or pyrolyzed solid) that the fire becomes underventilated. Temperatures are usually high and such fires tend to be "flashed over" [10]. The findings of this project will ultimately be incorporated into *Hazard I*.

Smoke inhalation has often been cited as the leading cause of death in fires. Over the years the central role of carbon monoxide in these deaths has been documented. Harwood and Hall [2] have considered the question: "What kills in fires: Smoke Inhalation or Burns?" Based on an analysis of death certificates in the United States, they concluded that the ratio of smoke inhalation to burns as the cause of death is roughly 2:1. For fires where death occurs beyond the room of origin for the fire, the ratio is even higher. Ominously, they concluded that the fraction of deaths due to smoke inhalation is increasing as time passes.

The toxicity of CO for humans is a result of the strong, long-lived bond which forms between the CO molecule and the hemoglobin in blood cells. [11] The complex of CO and hemoglobin is referred to carboxyhemoglobin. When CO enters the lungs it is absorbed into the bloodstream where it ties up the hemoglobin in red blood cells which would otherwise carry oxygen to the body. When the percentage of the resulting carboxyhemoglobin is high, insufficient oxygen is available for the body and the victim first becomes drowsy, then incapacitated, and if the level is sufficiently high, ultimately suffocates to death. Harland and Anderson considered the carboxyhemoglobin levels of fire victims. [1] These authors concluded that roughly 55% of fire victims have fatal carboxyhemoglobin levels (> 50%) and that an additional 13% of the remaining victims had levels sufficient to cause some incapacitation. Roughly two thirds of the fire deaths can therefore be attributed to the effects of CO.

Taken together, the findings of the two studies [1],[2] cited above suggest that roughly two thirds of all fire deaths are the result of CO generation in enclosure fires (approximately 2520 of 3765 deaths per year in the United States [12]). Babrauskas et al. have considered the types of fires which are most likely to result in CO-induced deaths. [3] They conclude that the most important type of fire by far is a post-flashover fire in which products of combustion are transported away from the room of fire origin. Unfortunately, it is this scenario for which the least information is available concerning the levels and mechanisms for the generation of CO.

Until recently, no full-scale experiments had been reported which were designed to investigate systematically the formation of CO in and transport of CO from room fires. On the other hand, concentrations of CO and other flame gases were routinely recorded, though without a systematic protocol, during full-scale fire tests. Mulholland of BFRL reviewed a number of full-scale fire tests performed at BFRL and reached conclusions concerning the formation of CO. [13] In a fully ventilated fire prior to flashover, the generation of CO is minimal and its toxicity is unlikely to be important. When a fire grows and achieves flashover there is a rapid increase in CO generation rates, and life-threatening concentrations of flame gases are quickly achieved. Based on his review, Mulholland recommended as a zeroth-order approximation that a generation rate of CO equal to 0.3 g of CO produced per gram of fuel burned be adopted for underventilated fires.

In 1987 a fire in a duplex townhouse in Sharon, PA resulted in the deaths of three people. One of the victims was reported to have had an extraordinarily high carboxyhemoglobin level of 91%. [14] Investigators from BFRL visited the site shortly after the fire and were able to learn a great deal about the building and the fire behavior by inspection and discussions with firemen and the building owner. The investigators concluded that this fire was a classic example of the fire scenario described above. The decision was made to do a simulation of this fire in the NIST full-scale fire facility. The goal was to characterize the fire and to test BFRL models designed to predict the fire behavior. Reference [14] summarizes the findings of this test.

The fire in Sharon was primarily confined to a downstairs kitchen which was panelled with wood and had a cellulosic ceiling. The fuel load was simulated using $\approx$ 185 kg of wood

6

cribs and plywood. The victims were found in upstairs bedrooms. During the fire test measurements of CO were recorded at the doorway soffit leading from the room adjacent to the kitchen as well as upstairs in the bedrooms. CO concentrations as high as 8.5% (dry volume) were observed at the doorway of the room adjacent to the kitchen, and concentrations greater than 5% were observed in the upstairs bedrooms. Such levels of CO will cause death nearly instantaneously and are consistent with the victims succumbing to smoke inhalation. Mulholland reviewed the results of the Sharon fire test and concluded that there was rough agreement between the measured CO levels and his zeroth-order model for CO generation in full-scale fire tests. [15]

## III. HOOD EXPERIMENTS AND THE GLOBAL EQUIVALENCE RATIO CONCEPT

### A. Introduction

Due to the lack of a substantial data base for CO formation in full-scale enclosure fires and a lack of understanding of the physical processes responsible for the generation of CO, it has been impossible to provide an engineering correlation or fundamental approach for predicting CO formation for the fire scenario of most interest. Progress has been made as described in this report.

Very early in the planning stages of the project, attention was focused on the results of experiments which were being performed at Harvard University and the California Institute of Technology. These studies are referred to as "hood" experiments. These experiments were intended to be idealized analogs of the two-layer zone models being developed to model room fire behavior. Zone models generally divide a room fire into two sections--a relatively cool and mildly vitiated lower layer and a layer containing the combustion gases which is generally hot and becomes vitiated for underventilated burning. Of course, during the history of a room fire the relative sizes, temperatures, and compositions of the two layers change dramatically.

### B.    Summary of Experimental Findings

In the experiments at Harvard University and the California Institute of Technology hoods were placed above fires burning in open laboratories. The vertical position of the flames could be varied such that for fully ventilated flames the combusting region actually extended into the hood. Combustion gases were trapped in the hood, which eventually filled up, forming an upper layer above the fire which was hotter than the ambient surroundings. For underventilated burning the flames would be quenched and products of incomplete combustion would be formed and trapped in the hood. Experiments showed that the gases in the upper layer, away from the fire plume, were well mixed. By varying the burner size and the separation of the fire source and the base of the upper layer, it was possible to change systematically the mass ratio of combustion gases in the hood derived from fuel and from entrained air. Generally, the fires were allowed to burn long enough for the temperature and composition of the upper layer to attain steady-state behavior.

The simple configuration and steady-state nature of the hood experiments allowed detailed measurements of species concentrations in the upper layer. From these measurements it was possible to derive the ratio of combustion product mass introduced from the fuel to the mass introduced from entrained air. In combustion science it is customary to divide this mass ratio by that required for complete burning of the fuel to fully oxidized products (for typical hydrocarbons the products are water and carbon dioxide ($CO_2$)). The resulting ratio is referred to as the equivalence ratio and is often denoted by the Greek letter phi ($\phi$). For lean mixtures $\phi$ is less than one, for stoichiometric mixtures $\phi$ equals one, and for rich mixtures $\phi$ is greater than one. For the hood experiments it is possible to define several different phi. Here we will consider two. The plume equivalence ratio ($\phi_p$) will be defined as the fuel mass flow rate divided by the air mass entrainment rate into the plume below the layer normalized by the stoichiometric ratio for the fuel. The global equivalence ratio (GER, denoted as $\phi_g$) will refer to the ratio of the mass of gas in the upper layer derived from the fuel divided by that introduced from air normalized by the stoichiometric ratio. Note that for steady-state cases, where no air or fuel enter the upper layer except by way of the fire plume, $\phi_p$ and $\phi_g$ are identical.

Beyler at Harvard University was the first to attempt to correlate his measurements of combustion gas concentrations in the upper layer with the GER. [16],[17],[18] Values of $\phi_p$ were determined by measuring the total gas flow into the hood (by balancing extraction of a known amount of gas from the hood to give a constant layer height) and using the measured fuel flow rate and compositions of the upper layer. Figures 1 and 2, taken from [16], are a schematic for the cylindrical hood and the overall experimental

9

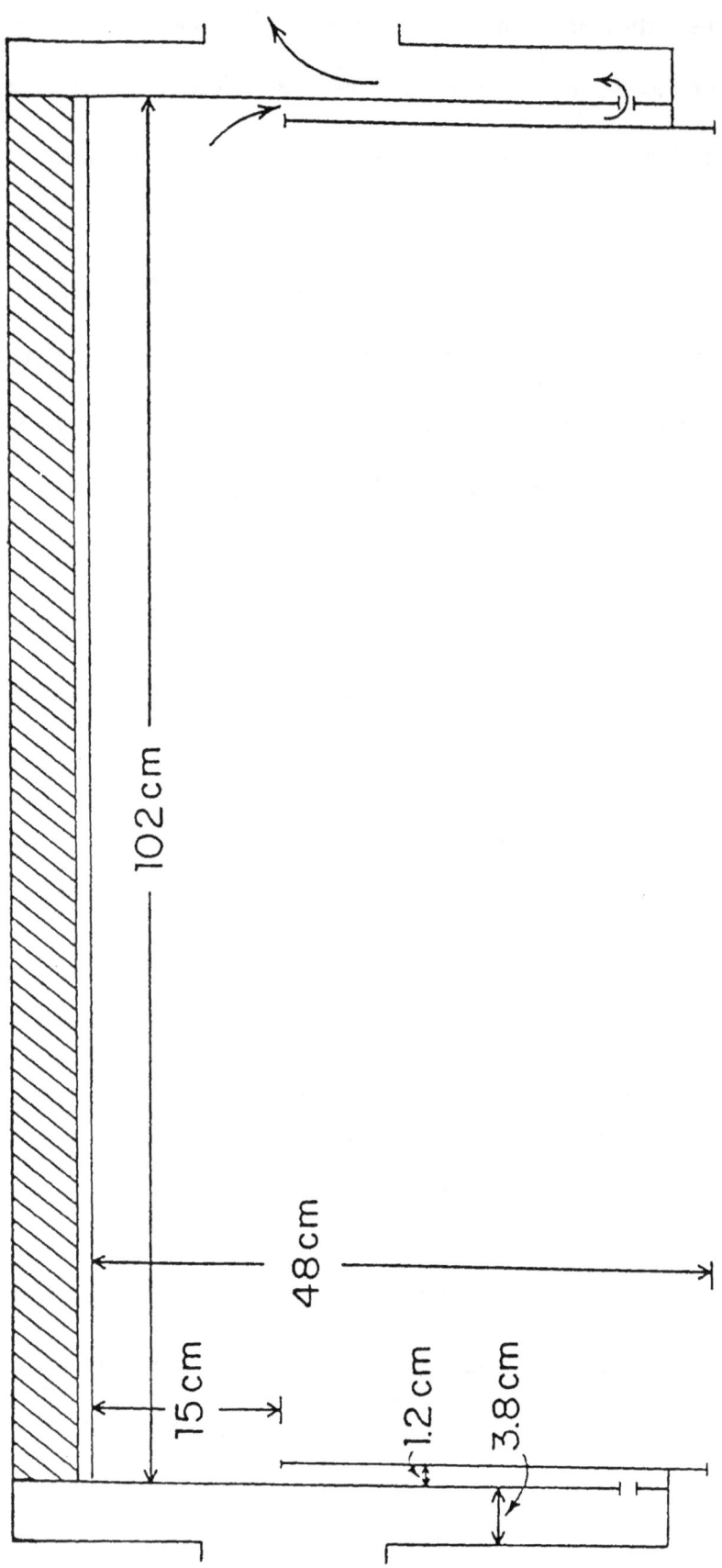

Figure 1.    Schematic of the cylindrical hood used by Beyler to investigate combustion gas generation by fires in a two-layer environment.  Figure reproduced from [16].

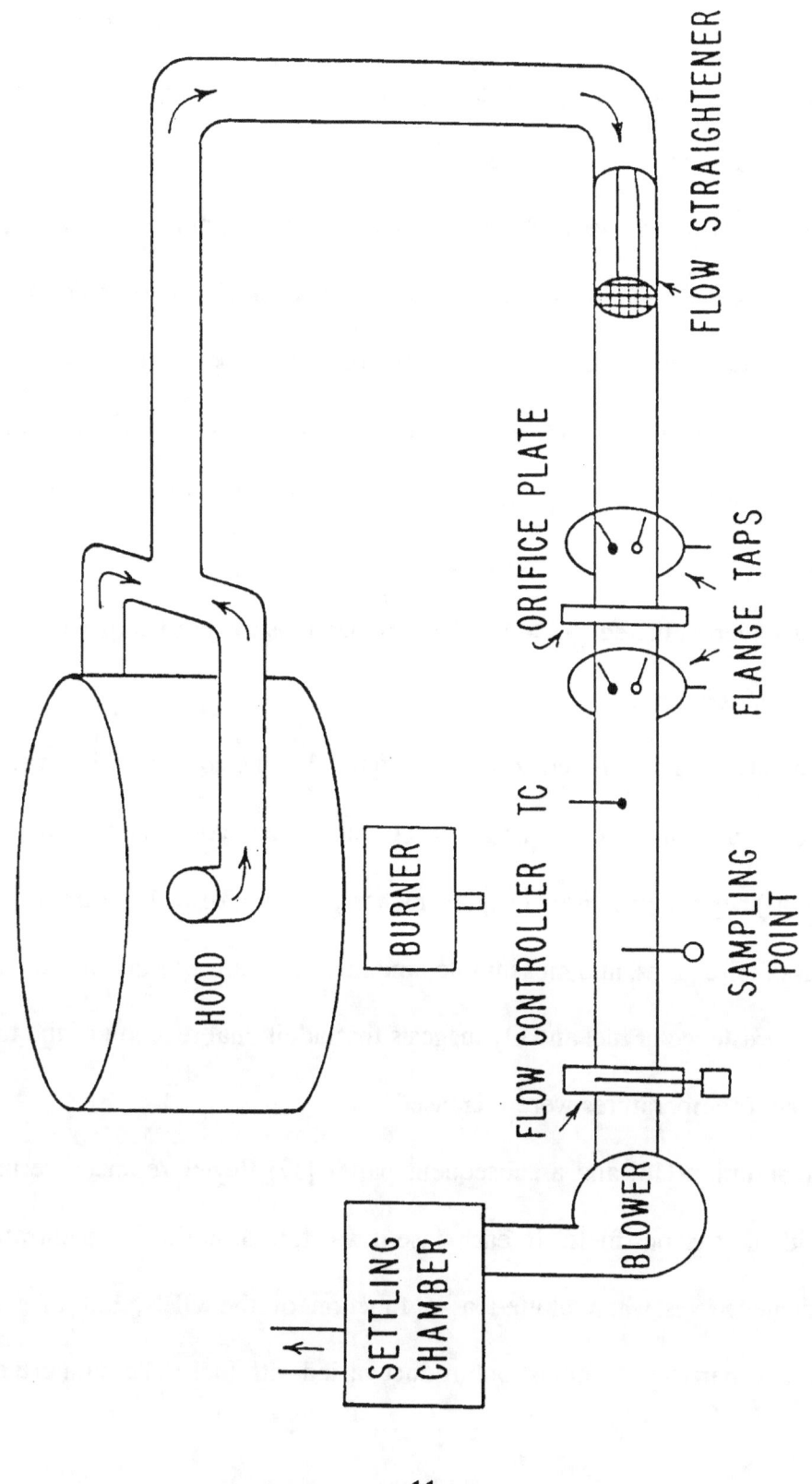

Figure 2. Schematic drawing of the hood and exhaust systems used by Beyler to investigate the formation of combustion gases for fires burning in a two-layer environment. Figure taken from [16].

configuration . Beyler found that the bottom of his layer was not well defined due to mixing between the combustion gases and the laboratory air. He also noted that flaming occurred along the bottom of the layer for $\phi_p$ in the range of 1.4-1.8. This layer interface burning limited the highest $\phi_p$ which could be investigated. [16],[19]

Since the experiments were done for steady-state conditions, to a very good approximation, $\phi_p$ is equal to $\phi_g$. Beyler found an excellent correlation of major gas species with $\phi_p$ which was independent of burner size, upper-layer temperature (range of 470-800 K), and separation of the upper layer and fire base. Figure 3 shows the correlations he observed for major gas species using propane as fuel. [16] Significantly, similar to enclosure fires, CO levels are low when the fire is overventilated ($\phi_p < 1$) and rapidly increase when the fire becomes underventilated ($\phi_p > 1$). The implications for predicting CO generation in enclosure fires are obvious.

Several points should be noted concerning figure 3. The concentration of CO starts to increase for $\phi_p > 0.5$ while the other products of incomplete combustion (hydrogen and total hydrocarbons) begin to increase at slightly higher $\phi_p$. For a $\phi_p$ of 1 there is roughly 2% $O_2$ remaining in the fire gases, and measurable amounts of $O_2$ are present at the highest $\phi_p$ observed. The coexistence of fuel and $O_2$ suggests that additional reaction of the fire gases would be possible if temperatures were increased.

In his dissertation [16] and a subsequent paper [17] Beyler reported results for a number of liquid and gaseous fuels. In each case it was found that the concentration data collapsed to single curves when plotted as a function of the GER, but that absolute concentrations for a particular combustion product varied with fuel. The data are reported

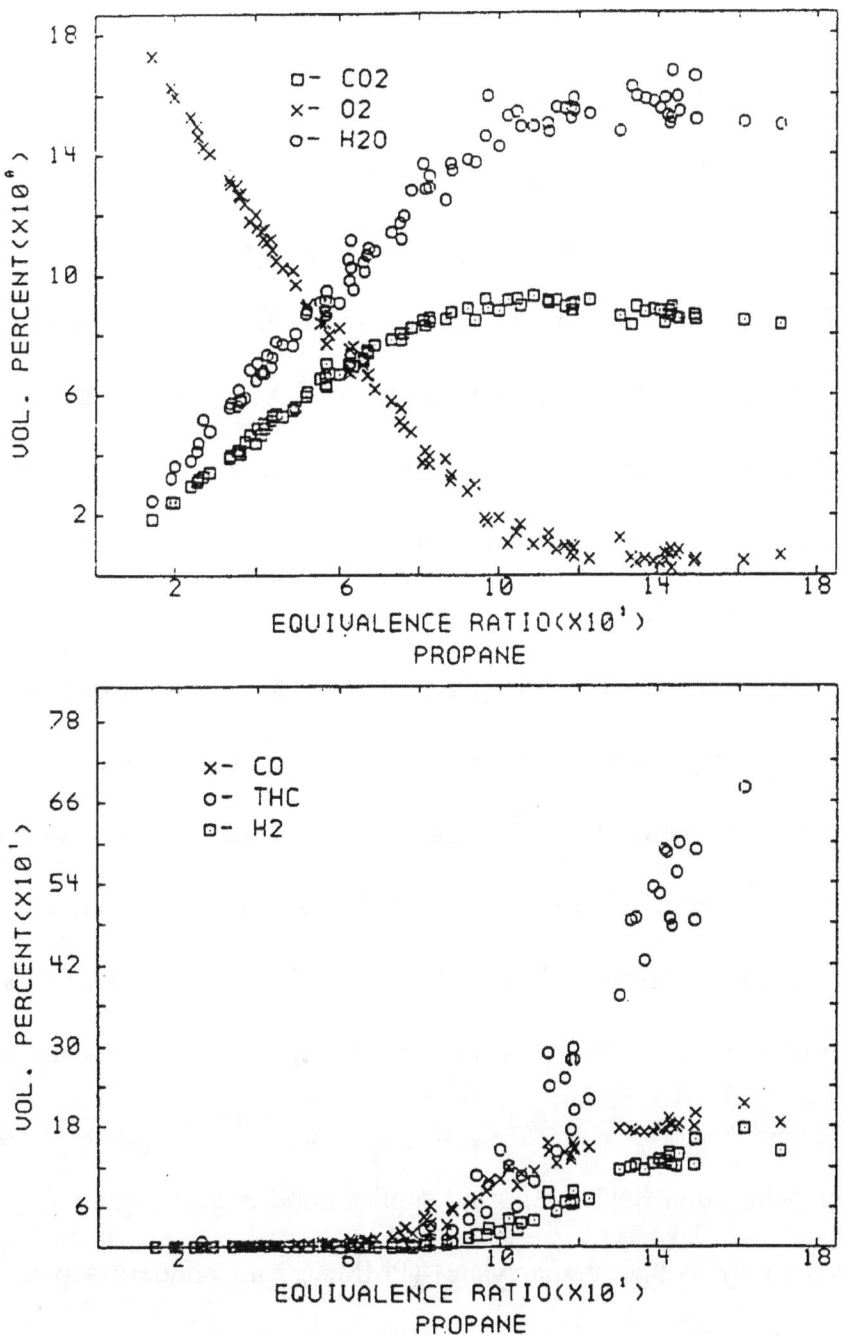

Figure 3.    Measured concentrations (in volume percent) of carbon dioxide, oxygen, water, carbon monoxide, total hydrocarbons, and hydrogen observed in a hood located above a propane flame are plotted as a function of $\phi_p$.  Figure is reproduced from Beyler [16].

in two forms: as volume percentages and as normalized yields defined to be the ratio of the actual and maximum theoretical production rates.

Beyler concluded that CO production could be characterized by two constants--one for lean ($\phi_p < 0.7$) and one for rich ($\phi_p > 1.2$) conditions with a transition region centered about $\phi_p = 1$. All fuels, with the exception of toluene which generated high CO concentrations for overventilated burning, resulted in low CO concentrations in the hood (much less than 1%) for $\phi_p < 0.7$. Considerably higher CO concentrations were generated for $\phi_p > 1.2$. Table 1 summarizes the findings for CO in terms of concentrations and yields observed for rich conditions. For later comparison purposes, Beyler's normalized yields have been converted to absolute yields.

Some interesting conclusions concerning fuel effects can be drawn from the results in table 1. Partially oxygenated fuels such as alcohols and ketones lead to higher concentrations (in volume percent) of CO than alkanes or alkenes. The lowest concentration of CO was observed for toluene. Beyler attributes this observation to the thermal stability of this molecule. Presumably, due to the stability of the fuel, CO successfully competes with it for the free radicals which drive the combustion process and is therefore more efficiently oxidized.

In a later publication Beyler reported similar hood measurements for three solid fuels: polyethylene, poly(methyl methacrylate) (PMMA), and ponderosa pine. [18] The results of these experiments have been included in table 1. The solid fuels appear to follow the same trends as the gases and liquids. Polyethylene, which contains only carbon and

14

Table 1. Beyler's Findings for CO Formation Under Rich Conditions

| Fuel | Formula | CO Vol. Percent | CO Yield (g/g) |
|---|---|---|---|
| Propane | $C_3H_8$ | 1.8 | 0.23 |
| Propene | $C_3H_6$ | 1.6 | 0.20 |
| Hexanes | $C_6H_{14}$ | 1.6 | 0.20 |
| Toluene | $C_7H_8$ | 0.7 | 0.11 |
| Methanol | $CH_3OH$ | 4.8 | 0.24 |
| Ethanol | $C_2H_5OH$ | 3.6 | 0.22 |
| Isopropanol | $C_3H_7OH$ | 2.4 | 0.17 |
| Acetone | $C_3H_6O$ | 4.4 | 0.30 |
| Polyethylene | $CH_2$ | 1.7 | 0.18 |
| Poly(methyl methacrylate) | $C_5H_8O_2$ | 3.0 | 0.19 |
| Pine | $C_{0.95}H_{2.4}O$ | 3.2 | 0.14 |

hydrogen, generates roughly the same concentration of CO as the alkanes and alkynes while the oxygen-containing fuels produce levels roughly comparable to the alcohols and ketones.

The major conclusions of Beyler's work concerning CO formation in the hood-type experiments are:

1. Major flame species including CO can be correlated in terms of $\phi_g$.

2. Relatively constant concentrations of CO are generated at low and high $\phi_g$.

3. The generation of CO under rich conditions is considerably greater than for fuel-lean conditions.

4. The concentrations of CO generated for rich conditions are fuel dependent, but can be correlated with fuel structure. Oxygen-containing fuels generate the highest CO levels while especially thermally stable fuels generate the lowest. Hydrocarbon fuels fall in the middle.

15

In an investigation concerned primarily with entrainment into a buoyancy-driven fire plume, Cetegen at the California Institute of Technology [20] used a hood system which was a significant modification of the hood design employed by Beyler [16]. Figure 4 shows a schematic for the experiment. It can be seen that instead of withdrawing gases directly from within the hood, as was done by Beyler, in order to control the position of the layer interface, the gases were allowed to spill out beneath the inner hood and into a second hood from which they were exhausted from the laboratory. As a result, the layer interface was located very close to the bottom of the first or catcher hood, which was a cube having 1.2 m sides. With this configuration the interface region is thinner and its location is much better defined than in the hood experiments of Beyler.

The fuel used by Cetegen was natural gas, which is primarily methane ($\approx$ 92%). Limited measurements were made of the concentrations of products (with water; i.e., dry measurements; and soot removed) within the upper layer above the fire. For rich conditions, upper-layer temperatures were on the order of 850 K. When plotted as a function of $\phi_p$, observed mole fractions (equivalent to volume fractions) of CO were negligible for $\phi_p < 0.6$, began to increase rapidly between $0.5 < \phi_p < 1.6$ and then remained nearly constant at $\approx$ 0.024 until the highest $\phi_p$ for which measurements were recorded ($\phi_p$ = 2.3). These findings are consistent with the observations of Beyler [16],[17] for a range of other fuels. Unfortunately, oxygen measurements were not reported. Interestingly, ignition of the layer interface was not observed until $\phi_p$ was greater than 2.5, suggesting that the bottom of the layer was better defined than in the experiments of Beyler [16],[19] where ignition was observed for $\phi_p > 1.8$.

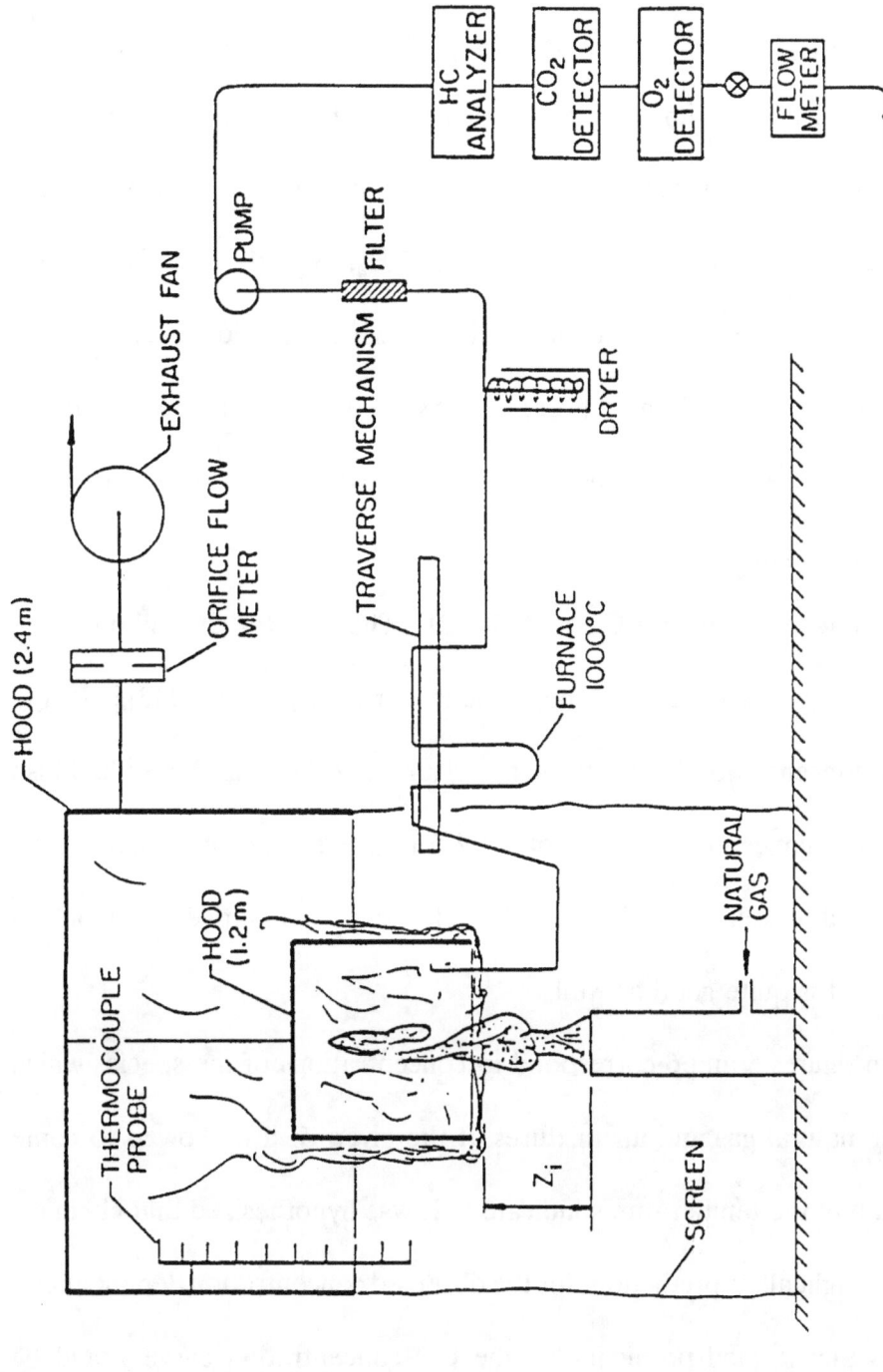

Figure 4. The experimental arrangement for the hood measurements of Cetegen [20]. Combustion gases are allowed to flow from beneath the first hood thus defining the layer location. Limited gas measurements were made for gases extracted directly from the layer. For these measurements the furnace shown in the figure was not operated.

17

In later work at the California Institute of Technology, Lim [21] used the same experimental facility with natural-gas fuel to make more careful measurements of concentrations in the upper layer. Concentrations of $CO_2$, CO, $O_2$, and $CH_4$ were measured for gases extracted from the hood. Similar to the earlier work, the species concentrations were independent of position for locations removed from the fire plume. By assuming mass balance, approximate concentrations for $CO_2$, CO, $O_2$, $CH_4$, $N_2$, $H_2$, and $H_2O$ were calculated for the layer and plotted as function of the $\phi_g$ which should equal $\phi_p$ since measurements were recorded for steady-state burning. Observed upper-layer temperatures were in the range of 450-850 K. Measurements were made while varying both the fuel source-layer interface distance and the fuel flow rate.

Figures 5 and 6 show the results for CO and $O_2$. [21] The data for CO fall on a well defined curve which has a similar shape to those found earlier by Beyler [16],[17] and Cetegen [20]. The CO concentration begins to increase for $\phi_g > 0.6$. As Beyler had also observed, there is a residual concentration of $O_2$ of 1-2% for $\phi_p > 1$. In summary, all of the hood experimental results discussed up to this point have the same qualitative behavior and the quantitative agreement is quite good as well.

The solid lines in figures 5 and 6 correspond to concentrations of the species which are calculated assuming natural gas and air mixtures at the given $\phi_g$ are allowed to come into chemical equilibrium at the temperatures indicated. It was hypothesized that chemical equilibrium calculations might allow predictions for the observed concentration dependencies on $\phi_g$. Figure 5 shows that a good prediction of the CO concentration curve would be obtained for a temperature between 750 K and 800 K. On the other hand, the chemical

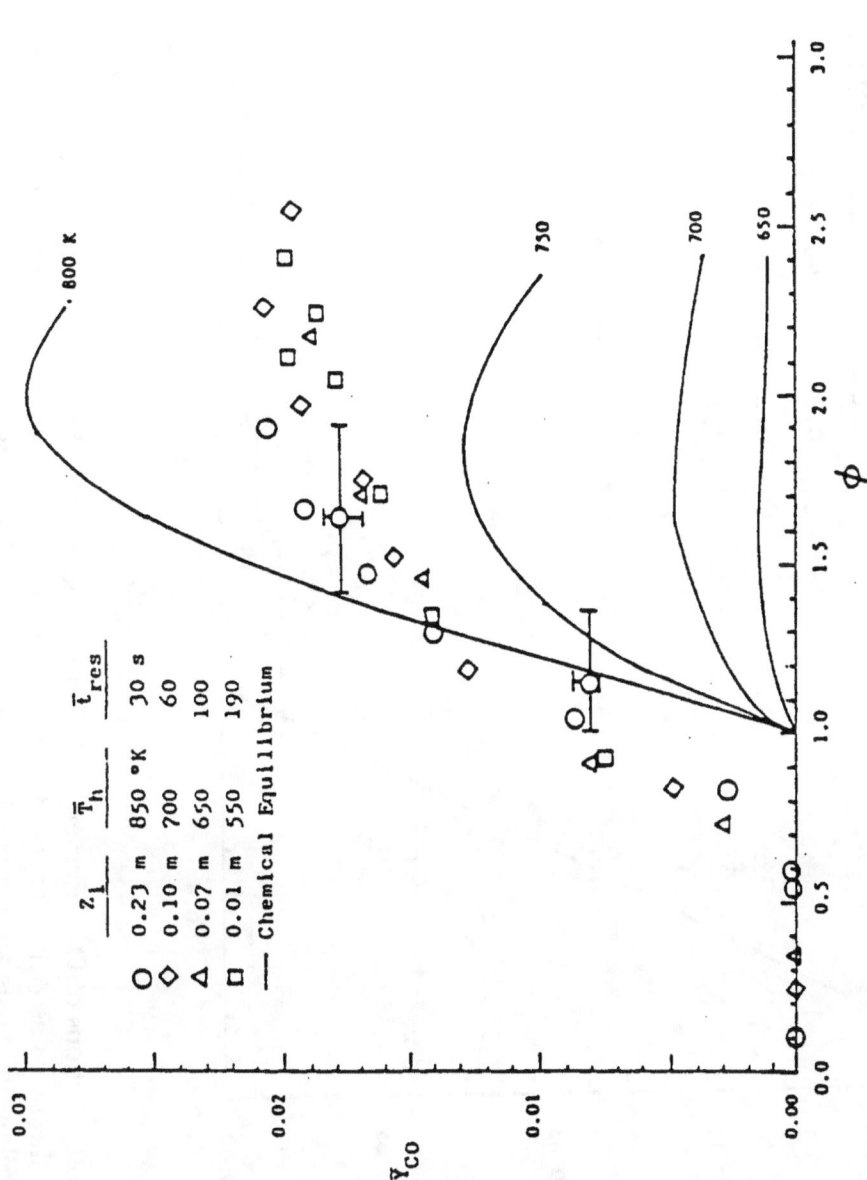

Figure 5.    Mole fractions of CO observed in the hood experiments of Lim [21] are plotted as a function of $\phi_p$ (which should be the same as $\phi_g$). The solid lines represent predicted CO concentrations assuming full equilibrium exists among combustion products for the temperatures indicated. Figure taken from [21].

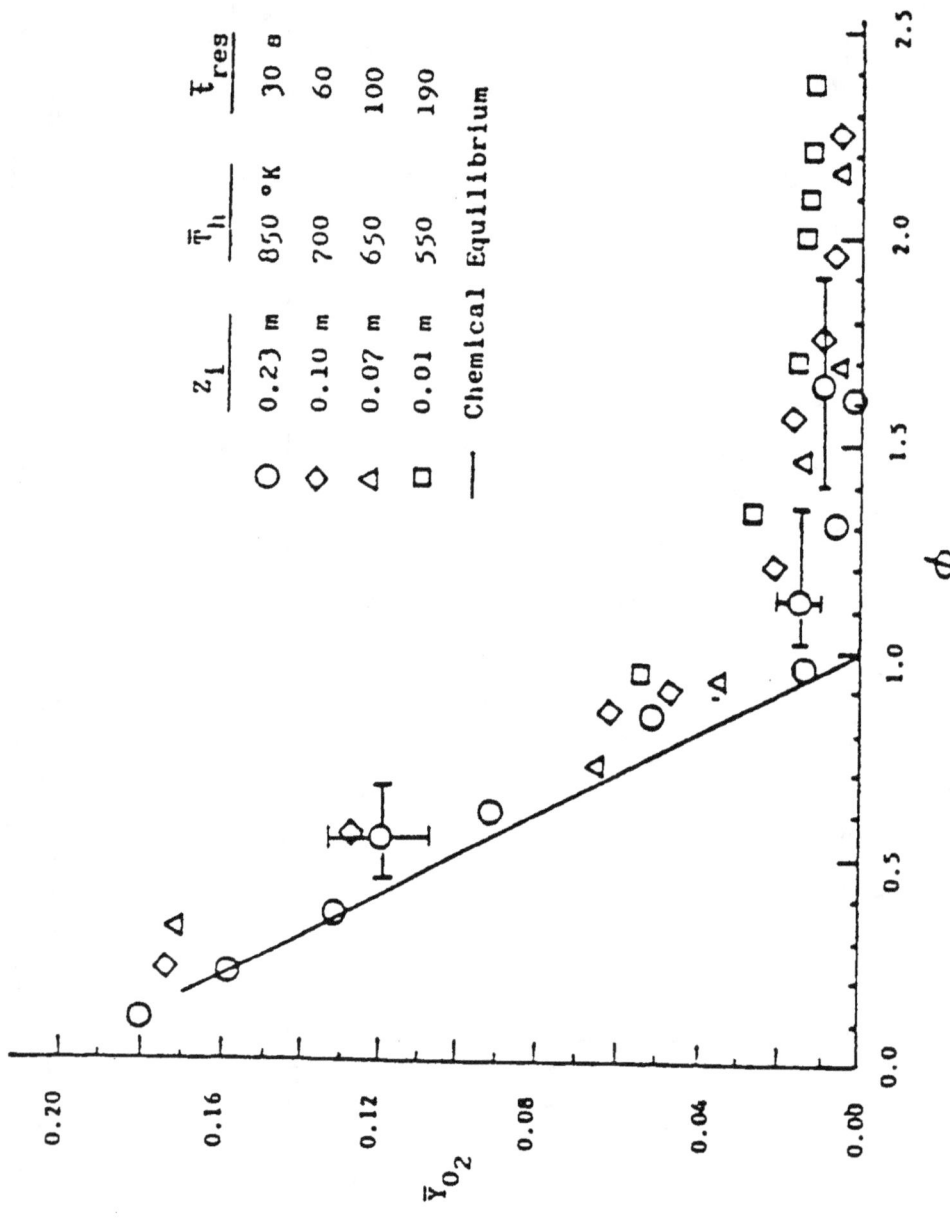

Figure 6. Mole fractions of $O_2$ observed in the hood experiments of Lim [21] are plotted as a function of $\phi_p$ (which should be the same as $\phi_g$). The solid line represents the predicted $O_2$ concentration dependence on $\phi_p$ assuming full equilibrium exists among combustion products. Figure taken from [21].

20

equilibrium calculations fail entirely to predict the residual oxygen concentrations observed for $\phi_g > 1$. The possibility of using equilibrium predictions to predict CO formation in enclosure fires will be assessed in Section IV.C.

The next improvement in the experimental procedure for the hood experiments was reported by Toner[22]. He replaced the individual gas analyzers used by earlier researchers with a gas chromatography system which allowed accurate concentration measurements of a large number of species in the upper layer contained within the hood constructed by Cetegen [20]. It was possible to record concentrations for a sufficient number of species so that by combining the measurements with appropriate conservation laws he was able to report all major species concentrations directly without making any assumptions concerning burning behavior.

Natural gas was the fuel. Checks showed the concentrations of combustion products in the upper layer to be uniform for positions far from the fire plume. Again the fuel flow rate and separation of the fuel source-interface height were varied. Temperatures in the upper layer having a range of 500-870 K were observed.

Figures 7-11 show the results of measurements for $CH_4$, $O_2$, $CO_2$, CO, and $H_2$. There are several points to note about these figures. All of the species concentrations are well correlated in terms of $\phi_g$. The only product of incomplete combustion observed for $\phi_g < 1$ is CO. $H_2$ and $CH_4$ only appear for $\phi_g > 1$. The CO mole fraction begins to increase for $\phi_g > 0.5$, being on the order of 1% for $\phi_g = 1$. CO concentrations appear to level off to a mole fraction of 0.019 for $\phi_g > 1.5$. These observations are consistent with the earlier work discussed above. A major difference from the earlier work is that oxygen

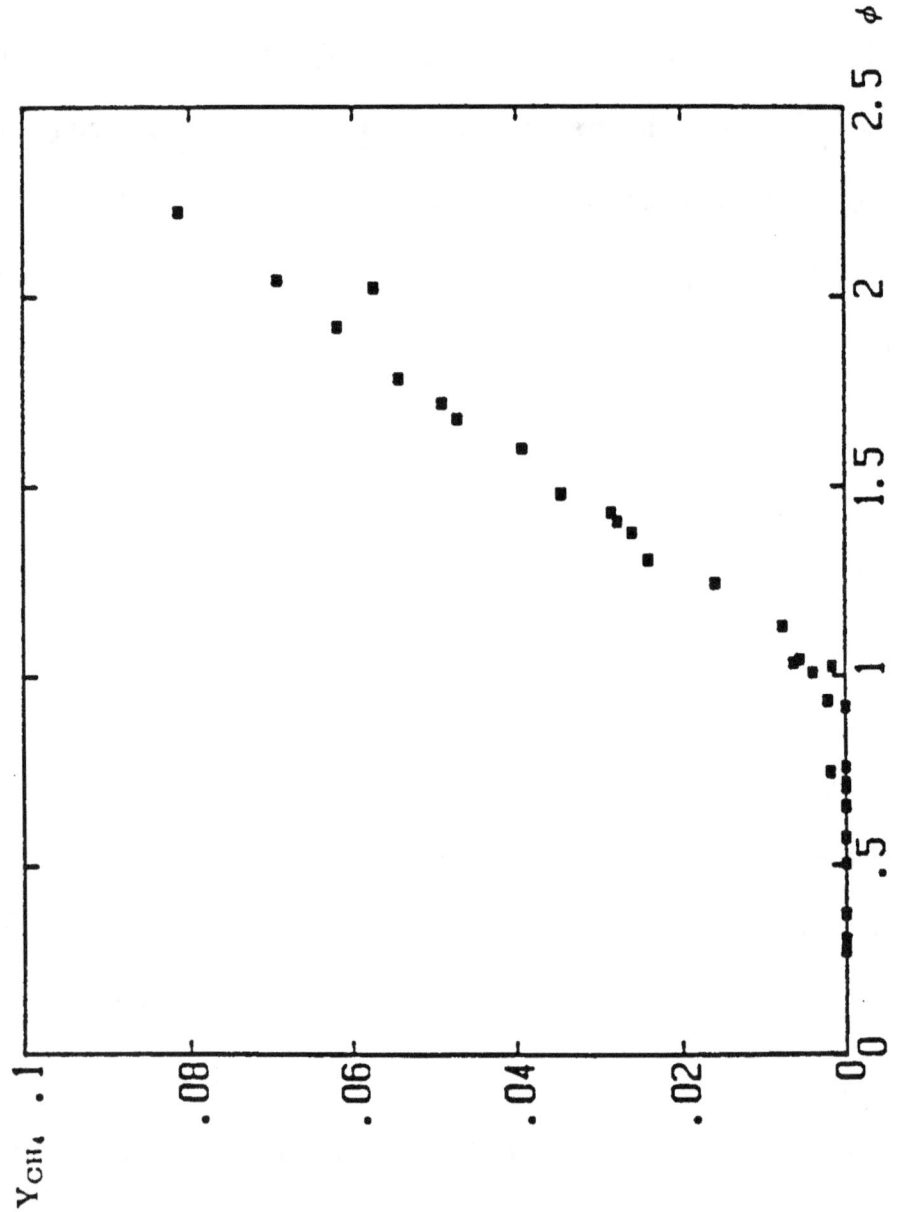

<u>Figure 7.</u>  Mole fractions of methane observed in the hood experiments of Toner [22] are plotted as a function of $\phi_p$ (which should be the same as $\phi_g$ for these experimental conditions).

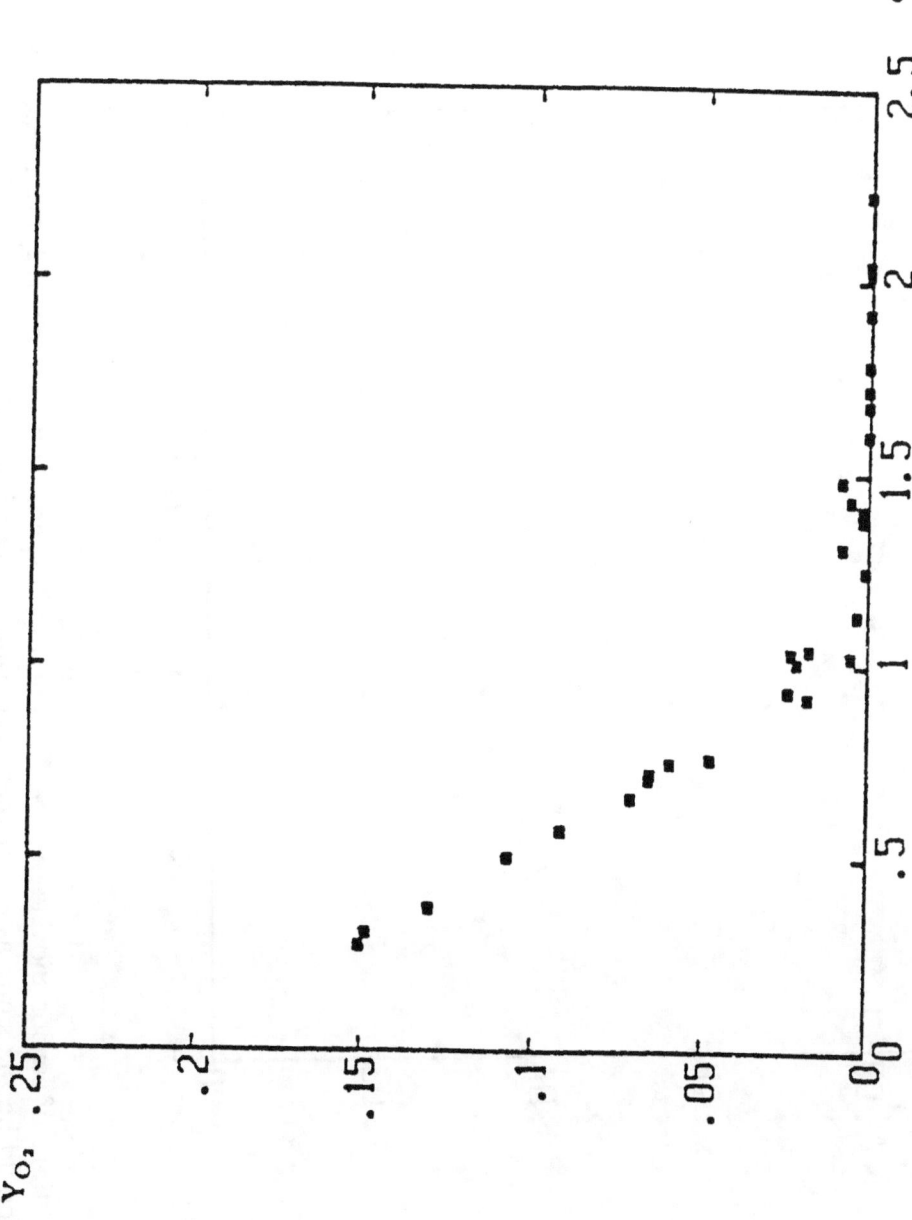

Figure 8. Mole fractions of oxygen observed in the hood experiments of Toner [22] are plotted as a function of $\phi_p$ (which should be the same as $\phi_g$ for these experimental conditions).

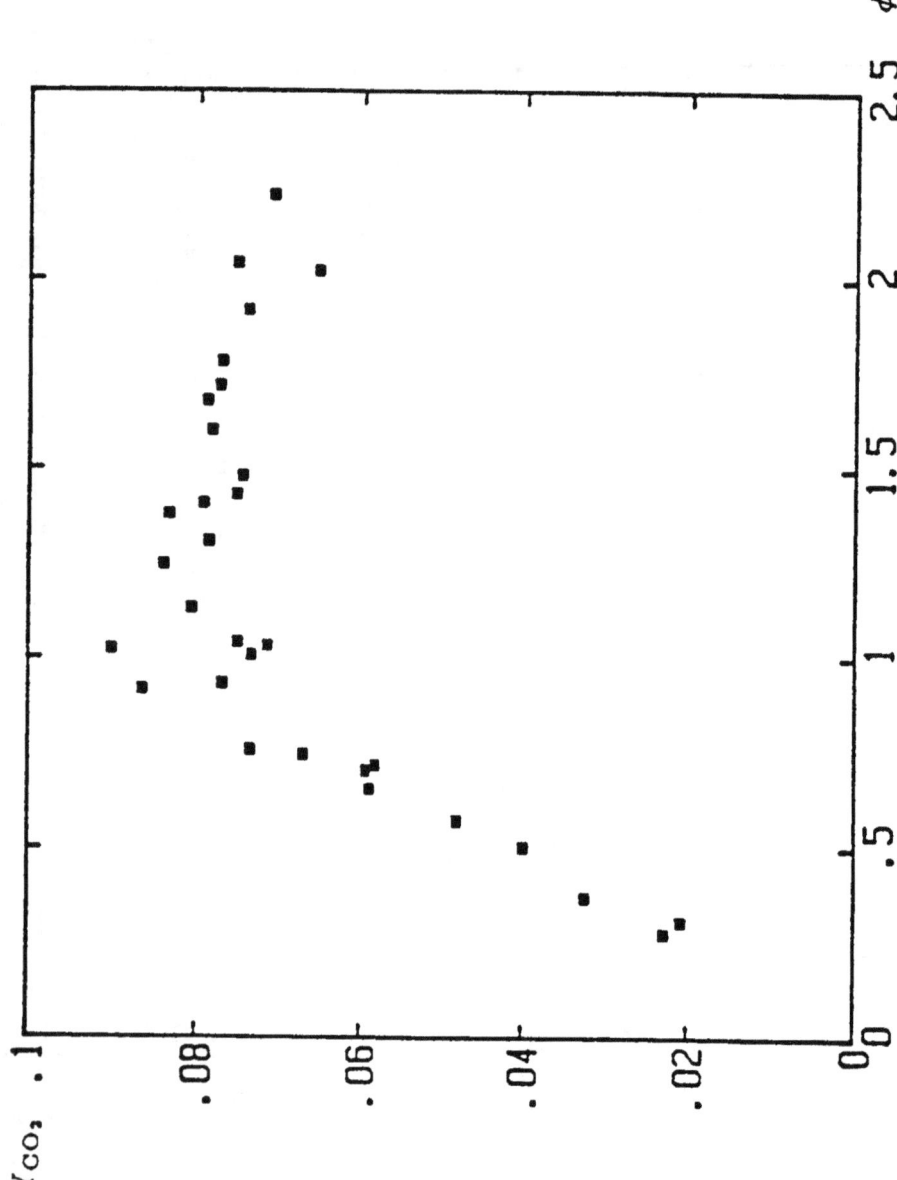

Figure 9. Mole fractions of carbon dioxide observed in the hood experiments of Toner [22] are plotted as a function of $\phi_p$ (which should be the same as $\phi_g$ for these experimental conditions).

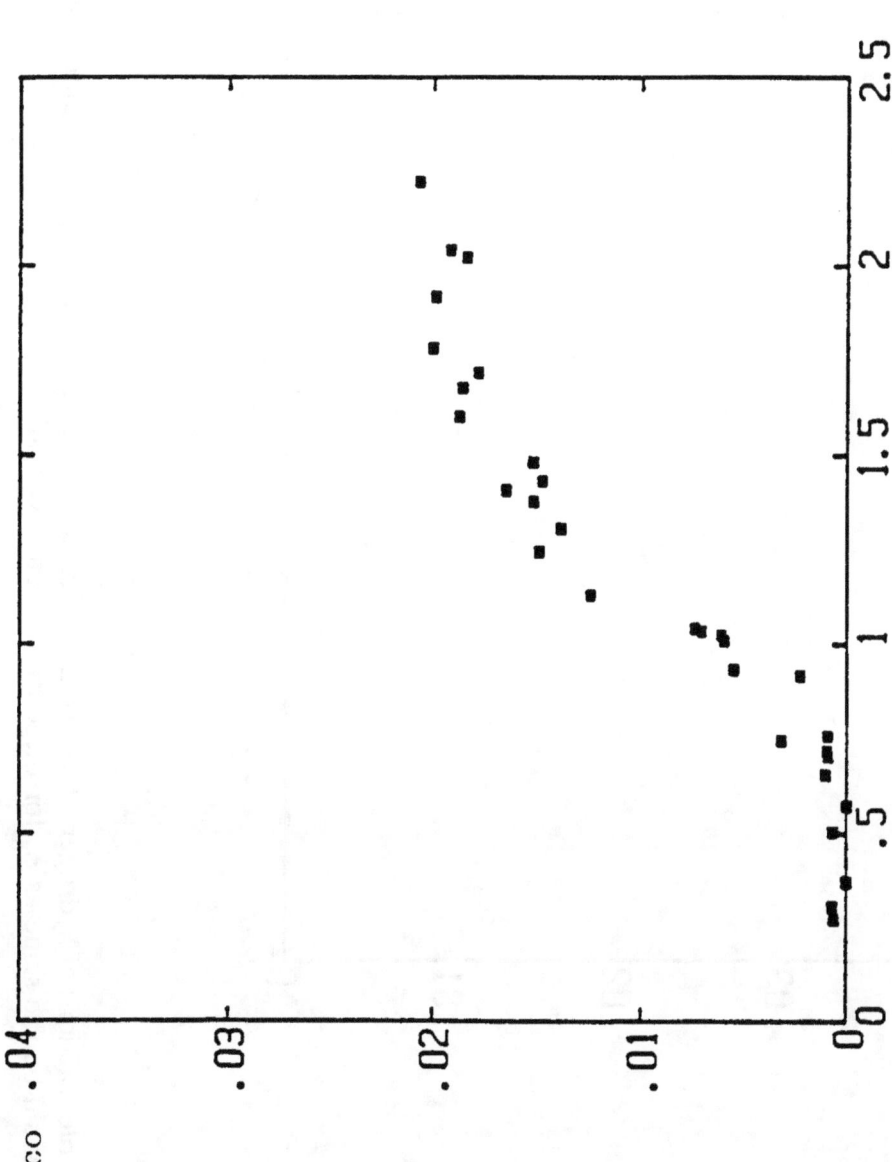

Figure 10. Mole fractions of carbon monoxide observed in the hood experiments of Toner [22] are plotted as a function of $\phi_p$ (which should be the same as $\phi_g$ for these experimental conditions).

25

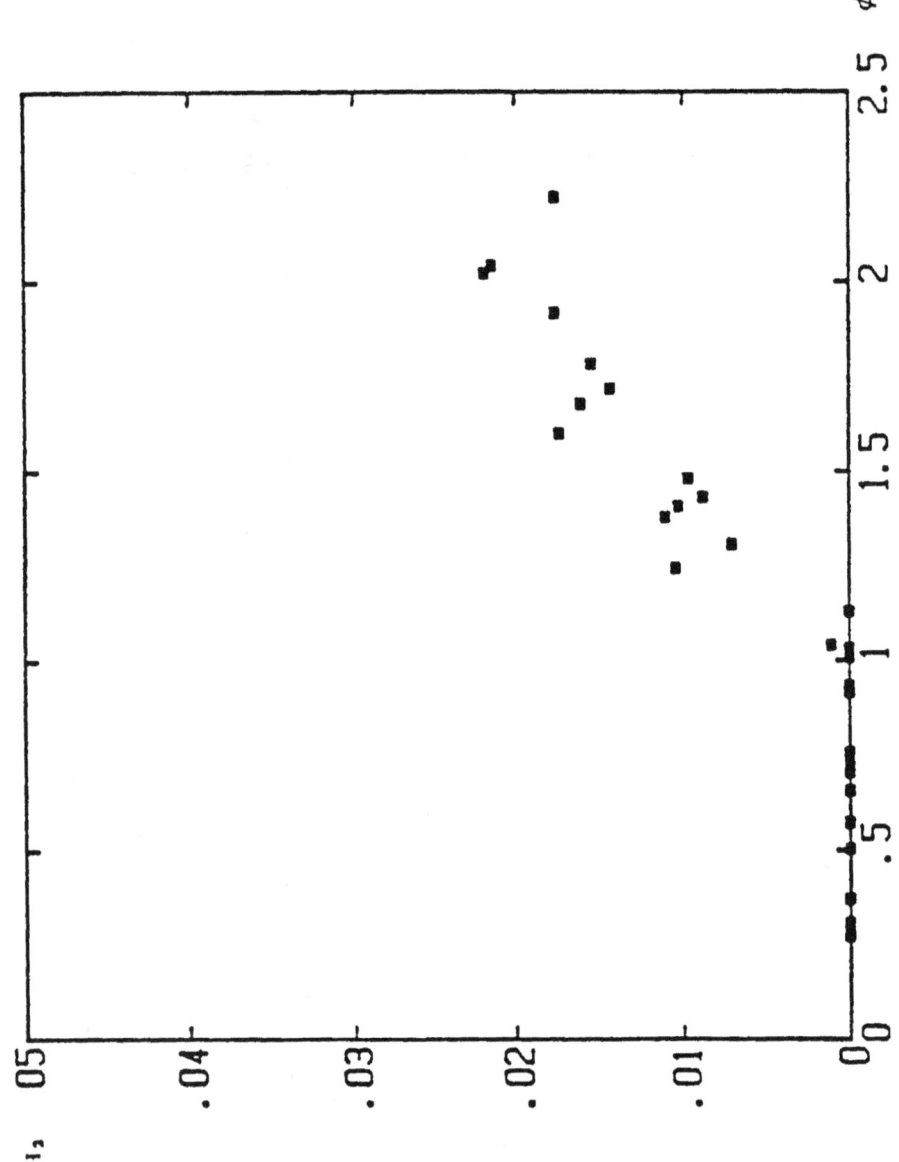

Figure 11.  Mole fractions of hydrogen observed in the hood experiments of Toner [22] are plotted as a function of $\phi_p$ (which should be the same as $\phi_g$ for these experimental conditions).

26

concentrations for $\phi_g > 1$ approach zero. Earlier experiments had suggested significant concentrations of $O_2$ remained for rich mixtures in the upper layers. This is the first indication that the correlations in terms of $\phi_g$ may depend on an uncharacterized (at the time this work appeared) variable.

The final series of hood experiments to be discussed were performed by Morehart, also working with Professors Zukoski and Kubota at the California Institute of Technology. [23],[24],[25] These measurements were made in a new hood facility, but utilized the same gas analysis procedures developed by Toner [22]. The new hood was considerably larger (1.8 m square x 1.2 m tall) than the earlier facility even though the upper layer was generated in a similar manner. A series of tubes with holes located within and near the top of the hood was incorporated such that additional gas could be injected directly into the upper layer at positions well removed from the fire plume. Figure 12 shows the arrangement of the tubes within the hood.

The new hood allowed $\phi_p$ and $\phi_g$ to be varied independently while maintaining a steady-state condition in the upper layer. For instance, by adding air to the upper layer it was possible to force $\phi_g$ to be lower than $\phi_p$. The experiment was designed to model the conditions expected in a developing room fire. In the early stages of such a fire there is sufficient oxygen available, and the upper layer which begins to form is fuel lean. As the fire grows, its oxygen (air) demands increase, and the fire plume entering the upper layer is richer than the combustion gases trapped above. Eventually the upper layer becomes rich enough (i.e., the oxygen concentration decreases sufficiently) to quench the fire plume in this layer, and concentrations of products of incomplete combustion build up. A pseudo-steady-

27

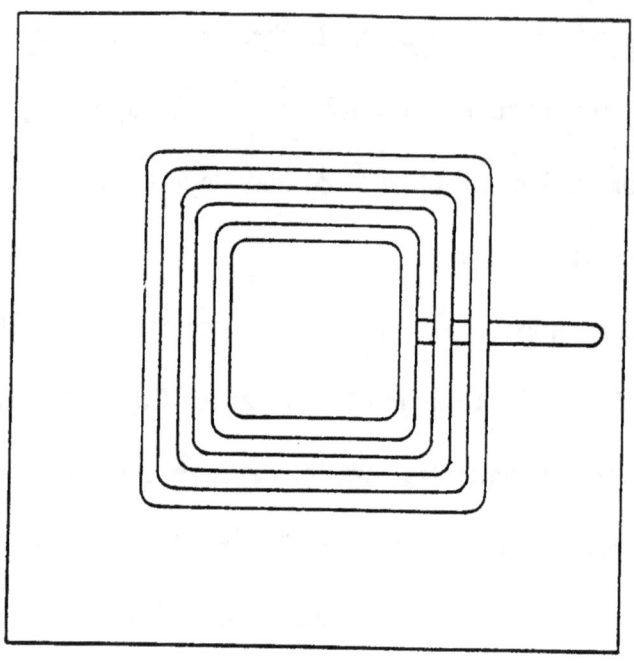

(a) Top view (through ceiling)

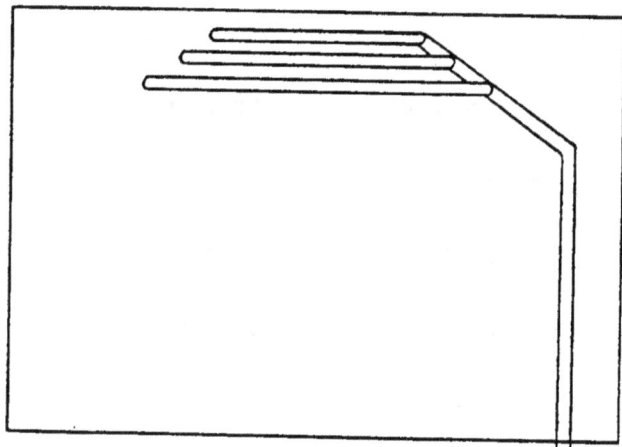

(b) Side view (showing injection orientation)

Figure 12.    Schematic of the tube arrangement used by Morehart to inject air directly into the upper layer above a natural gas fire. Added gas enters the layer through 365 1.6 mm holes spaced along the tubing. Figure taken from Morehart [23].

state burning may be reached in which $\phi_p$ and $\phi_g$ are the same (the condition modeled by the earlier experiments). A dying fire is expected to go through a reverse process. As the fire size decreases less oxygen is required, and $\phi_p$ becomes less than $\phi_g$. It would be possible to model this process in the hood experiments by adding fuel to the upper layer. Such measurements were not reported by Morehart.

The outline of a model for incorporating the findings of these new measurements into a description for a developing fire in an enclosure was provided by Morehart [23]. The model was not solved and no comparisons with actual fire behaviors were attempted. Cooper of BFRL has developed a different approach for incorporating the observations of the hood experiments into a zone model [26],[27],[28]. Neither of these models are discussed in this report.

The experimental approach used by Morehart [23] was to first make steady-state measurements without adding air to the upper layer (i.e., $\phi_g = \phi_p$). Values of $\phi_g$ and individual species concentrations were determined in the manner introduced by Toner [22]. This initial condition corresponds to the approach of the earlier hood experiments discussed above. Next, while maintaining the same fuel flow rate and fuel source-layer interface separation (which leaves $\phi_p$ unaltered), air was added to the upper layer thereby reducing $\phi_g$. Values of $\phi_g$ and hood concentrations were determined by repeating measurements of the major combustion gases for the new condition. This process was then repeated for several $\phi_g$. The procedure was repeated for several $\phi_p$.

Most of the experiments reported by Morehart used natural gas as fuel. Measurements at various locations within the hood showed that molecular concentrations were

29

uniform for positions outside of the fire plume, and that the temperature increased slightly from the bottom to the top of the enclosure. Hood temperatures for the results to be summarized here were in the 500-600 K range. As is discussed later in this section, due to the larger volume of the hood and differences in insulation, these temperatures are generally lower than observed in the earlier hood experiments.

Figures 13-19 show the results of measurements for $CH_4$, $O_2$, $CO_2$, $CO$, $H_2$, $C_2H_6$, and $C_2H_2$. [23] These figures contain a great deal of information which requires additional explanation. The open symbols are concentration measurements (in terms of mass fraction) for which no air has been added to the upper layer (i.e., $\phi_p = \phi_g$). Values of $\phi_p$ have been varied by changing the fuel flow rates and fuel source-layer interface separation. Each $\phi_p$ is represented by a different symbol. Experiments for which air has been added to the upper layer (thus lowering the $\phi_g$) are represented with corresponding solid symbols. The results of Toner [22] have been included on the same plot and are represented by the "+" symbol.

From figures 13-19 it is clear that the major species concentrations within the hood are well correlated when plotted as a function of $\phi_g$. The results for $\phi_g < \phi_p$ and $\phi_g = \phi_p$ lie on the same curve. The remarkable conclusion which is reached is that the concentrations of combustion gases in the upper layer depend only on the value of $\phi_g$ suggesting that the generation rates for the chemical species only depend on this variable. The importance of this observation is great because it suggests that steady-state measurements of this type could be used to predict instantaneous concentrations during a developing fire.

30

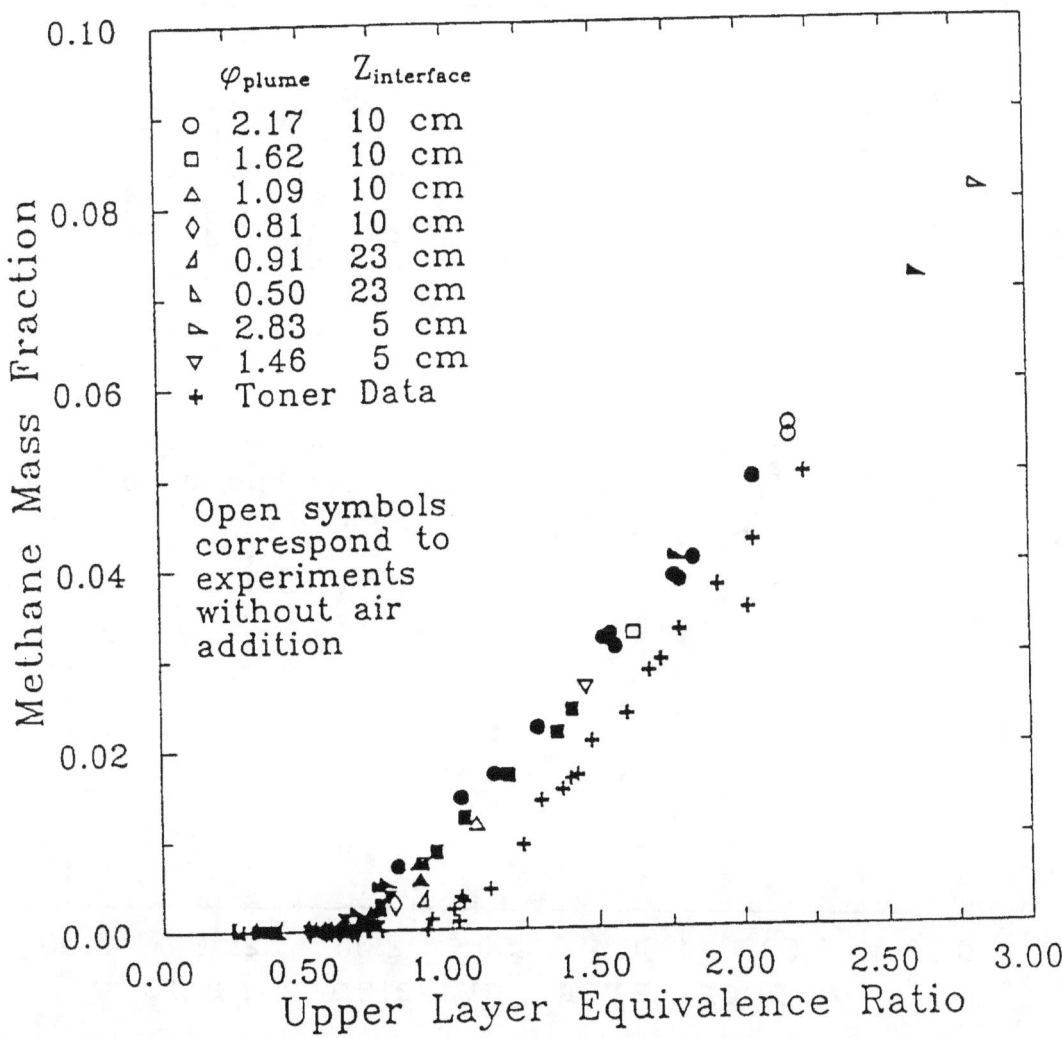

<u>Figure 13.</u>   Methane mass fractions observed by Morehart [23] in the combustion gases trapped in an upper layer above natural gas fires are plotted as a function of $\phi_g$. Filled symbols correspond to cases where $\phi_p \neq \phi_g$. The results of similar measurements by Toner [22] are included for comparison. Figure reproduced from [23].

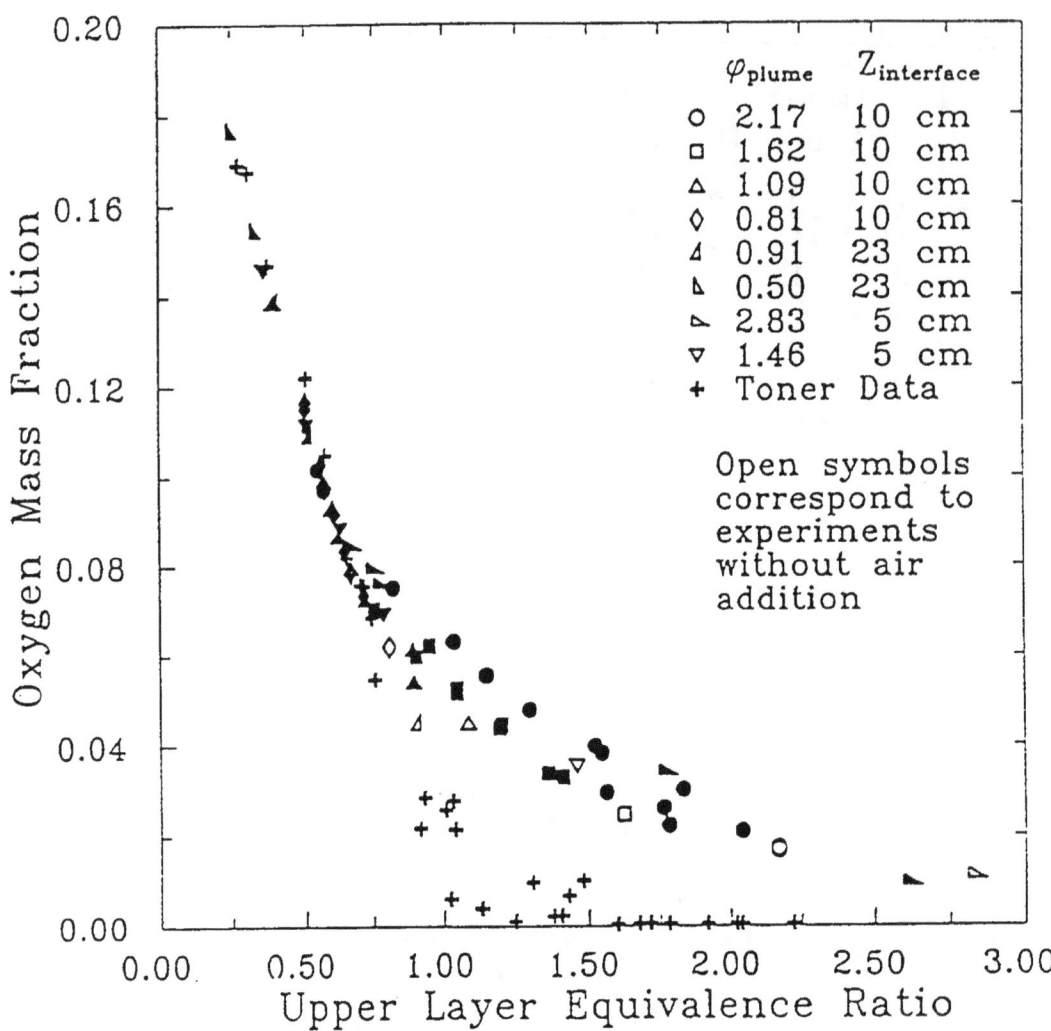

Figure 14. Oxygen mass fractions observed by Morehart [23] in the combustion gases trapped in an upper layer above natural gas fires are plotted as a function of $\phi_g$. Filled symbols correspond to cases where $\phi_p \neq \phi_g$. The results of similar measurements by Toner [22] are included for comparison. Figure reproduced from [23].

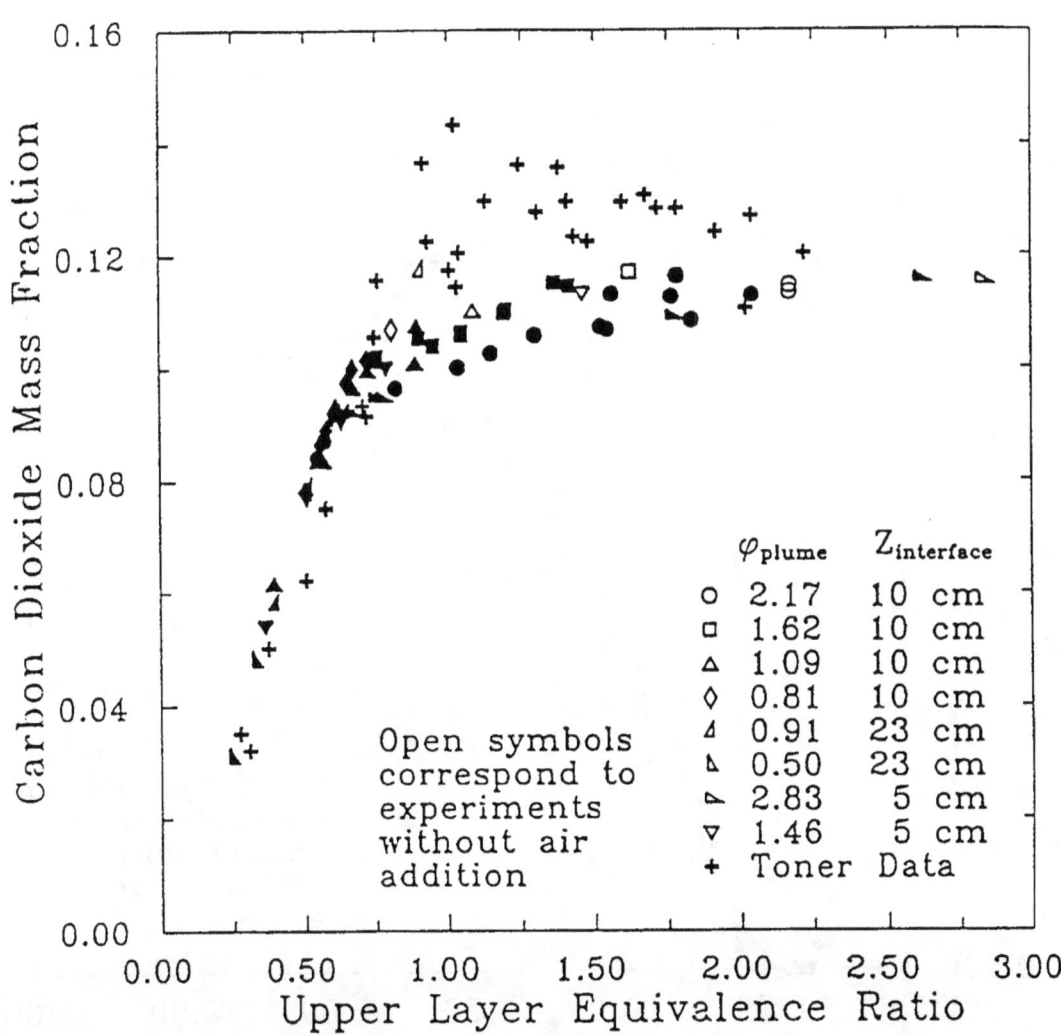

Figure 15. Carbon dioxide mass fractions observed by Morehart [23] in the combustion gases trapped in an upper layer above natural gas fires are plotted as a function of $\phi_g$. Filled symbols correspond to cases where $\phi_p \neq \phi_g$. The results of similar measurements by Toner [22] are included for comparison. Figure reproduced from [23].

33

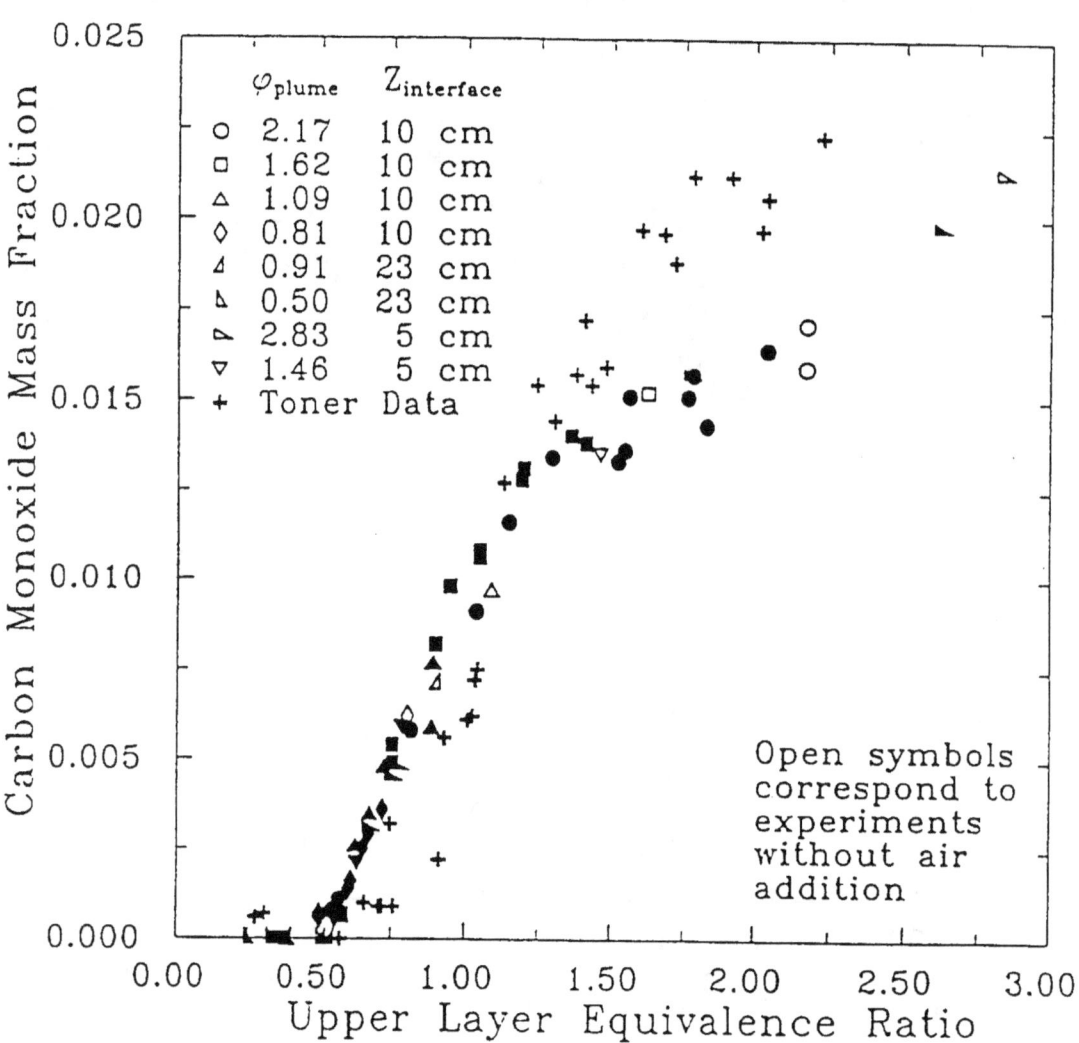

Figure 16.    Carbon monoxide mass fractions observed by Morehart [23] in the combustion gases trapped in an upper layer above natural gas fires are plotted as a function of $\phi_g$. Filled symbols correspond to cases where $\phi_p \neq \phi_g$. The results of similar measurements by Toner [22] are included for comparison. Figure reproduced from [23].

34

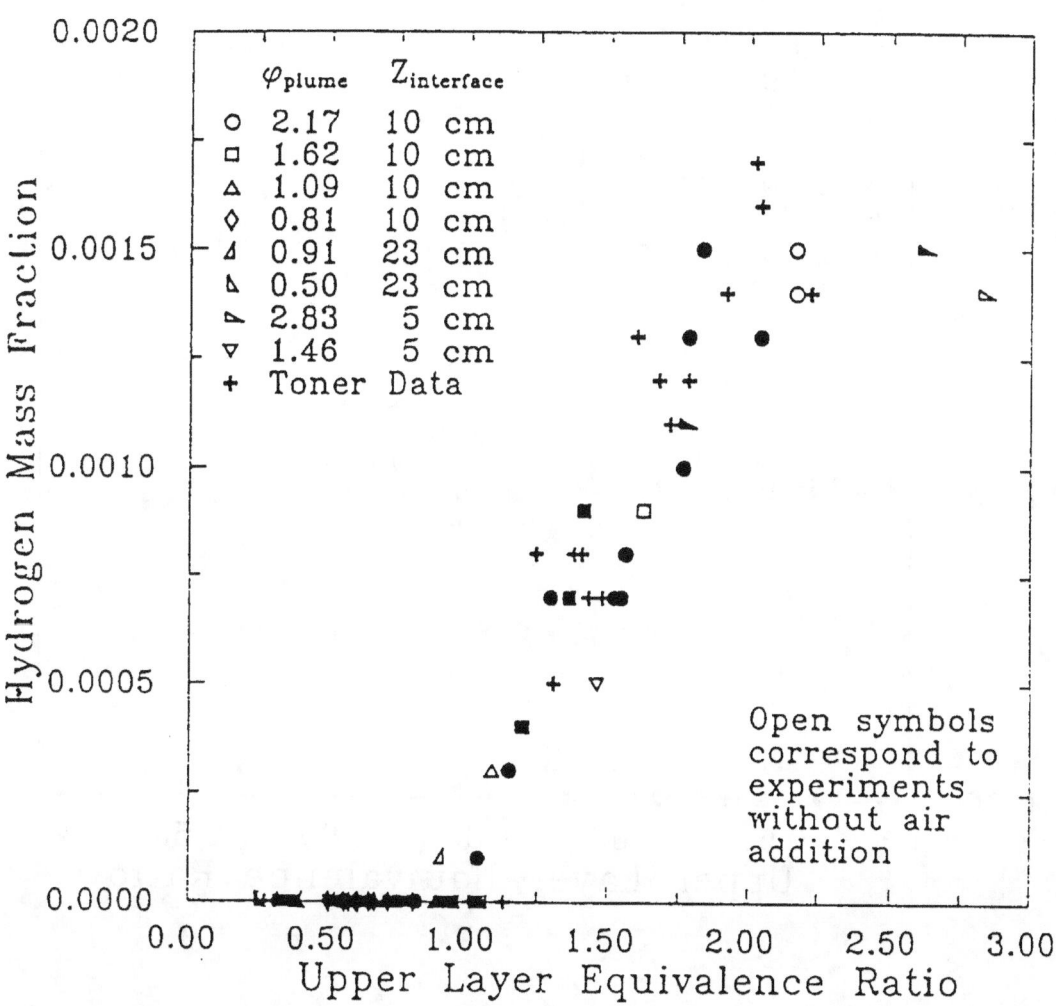

<u>Figure 17.</u>    Hydrogen mass fractions observed by Morehart [23] in the combustion gases
trapped in an upper layer above natural gas fires are plotted as a function of
$\phi_g$. Filled symbols correspond to cases where $\phi_p \neq \phi_g$. The results of similar
measurements by Toner [22] are included for comparison. Figure reproduced
from [23].

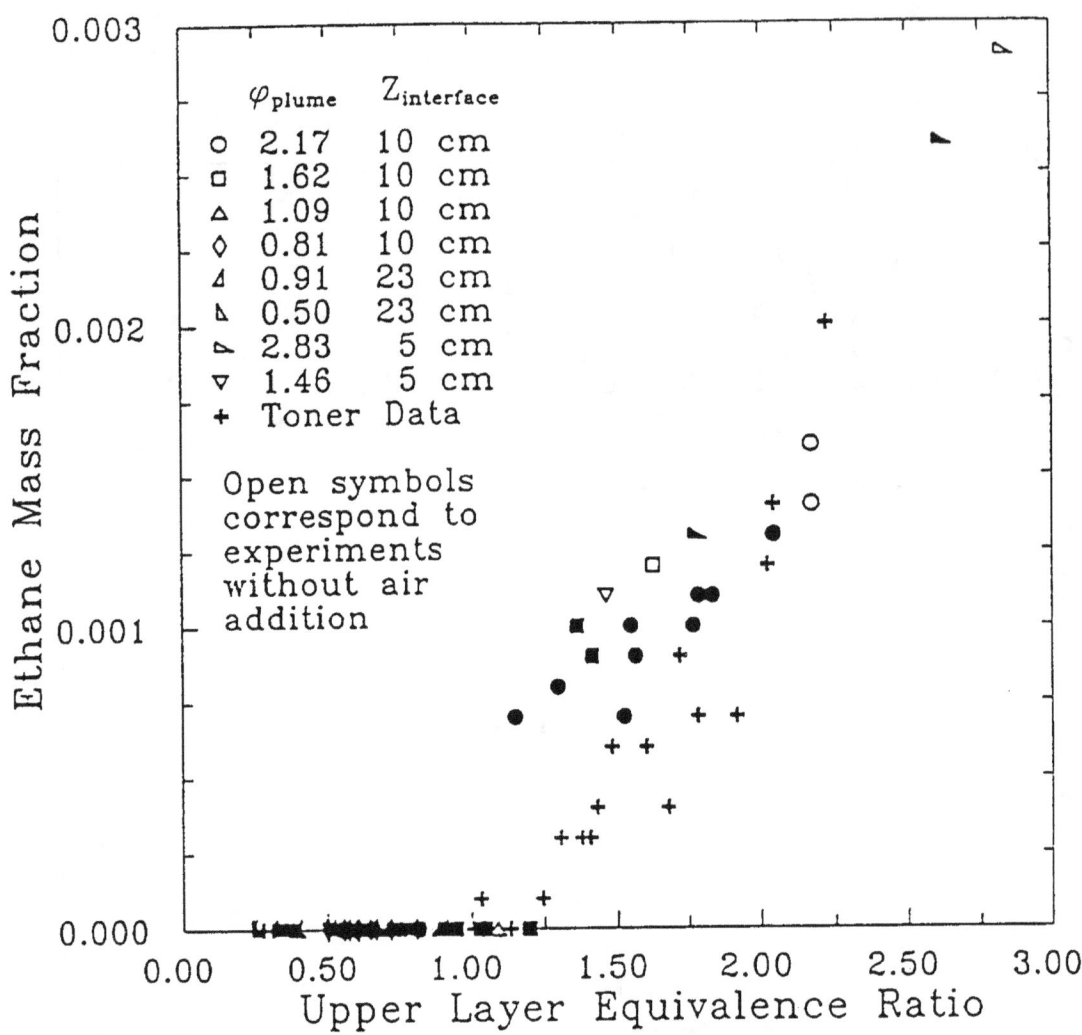

Figure 18. Ethane mass fractions observed by Morehart [23] in the combustion gases trapped in an upper layer above natural gas fires are plotted as a function of $\phi_g$. Filled symbols correspond to cases where $\phi_p \neq \phi_g$. The results of similar measurements by Toner [22] are included for comparison. Figure reproduced from [23].

36

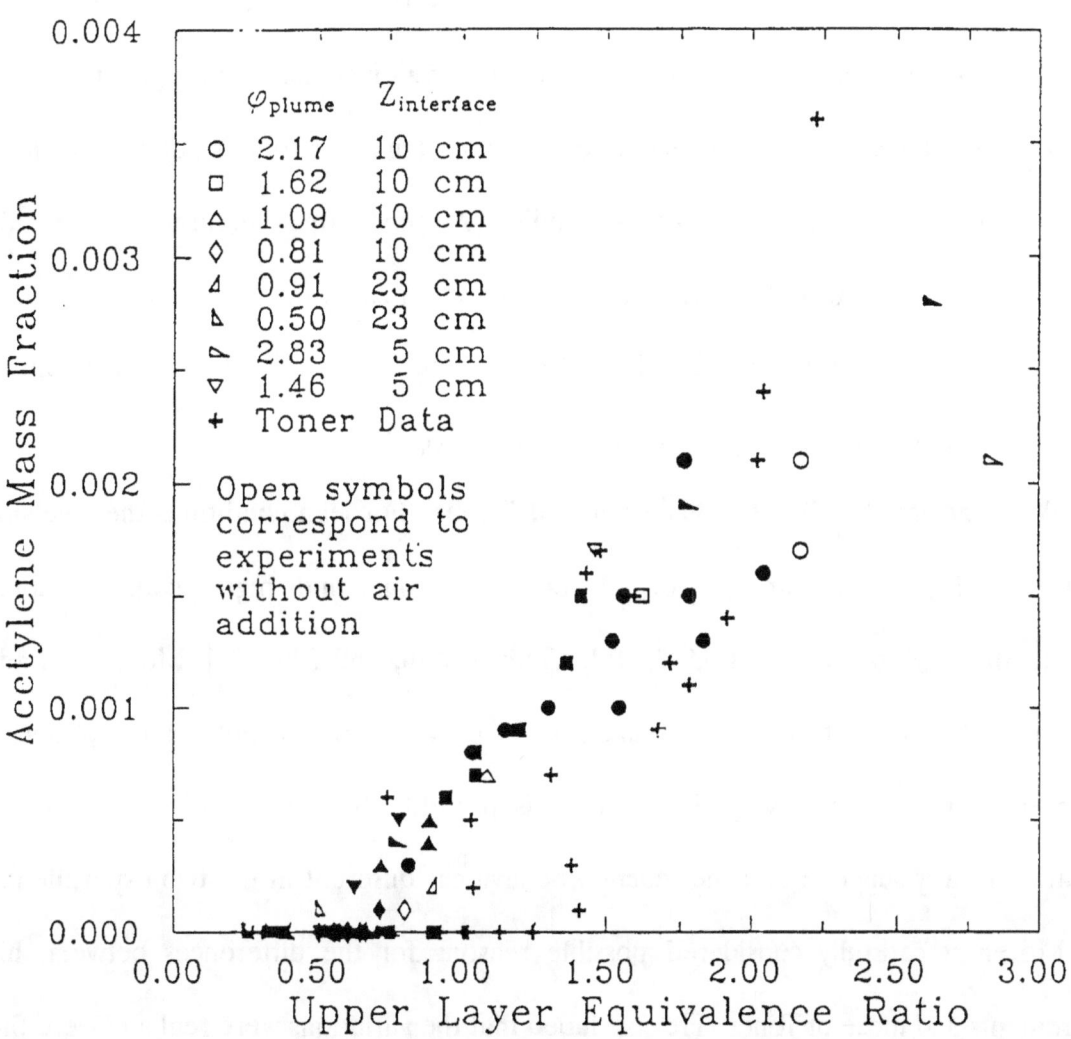

Figure 19. Acetylene mass fractions observed by Morehart [23] in the combustion gases trapped in an upper layer above natural gas fires are plotted as a function of $\phi_g$. Filled symbols correspond to cases where $\phi_p \neq \phi_g$. The results of similar measurements by Toner [22] are included for comparison. Figure reproduced from [23].

37

Unfortunately, Morehart's results also revealed a new uncertainty concerning the GER concept. Examination of figures 13-19 shows that there are significant systematic differences between the findings of Morehart [23] and Toner [22]. These are similar experiments using the same fuel. The differences are varied and complex. Consider the products of incomplete combustion. Both studies show CO increasing at roughly the same rate for $\phi_g > 0.5$, but CO concentrations for $\phi_g > 1.3$ are higher in Toner's hood. Acetylene shows a roughly similar behavior in the two studies. Morehart finds unburned $CH_4$ for $\phi_g > 0.5$ while it is only detected for $\phi_g > 1$ by Toner. In all cases the $CH_4$ concentration is lower in the Toner experiments. Finally, measurements for $H_2$ and $C_2H_6$ have similar behaviors in both cases, only beginning to increase for $\phi_g > 1$.

The behaviors for $CO_2$ and $O_2$ are also different. For lean conditions the levels of $CO_2$ observed by Morehart are greater, but for $\phi_g > 1$ Toner finds higher concentrations. Similar to the work of Beyler [16],[17],[18], Cetegen [20], and Lim [21], Morehart finds decreasing, but significant, concentrations of $O_2$ for $\phi_g > 1$, while, as noted above, Toner's measurements [22] indicate very little oxygen is present for rich conditions. This last observation clearly suggests that the reaction behavior is different in the two experiments.

Morehart carefully considered possible reasons for the differences between his measurements and those of Toner. He concluded that the variations were real and were the result of differences in an experimental parameter. He considered two possibilities: 1) insulation included within the Toner hood, and absent in the new hood, trapped soot and induced reaction; and 2) the higher temperatures observed in the earlier work led to different reaction behavior [23],[25].

Morehart tested both hypotheses experimentally. The addition of insulation material to the inside of his hood did not change the observed reaction products. Next he added insulation to the outside of his hood in order to raise the temperature. The insulation was added in stages which systematically raised the observed hood temperatures. As shown in figure 20, which compares results for $\phi_g = 1.45$, variations of mass fractions were observed in the upper layer as the temperature was changed. These measurements are compared with the results of Toner [22] in the figure. While not exact, the dependence on temperature agrees closely with Toner's measurements. These findings show conclusively that the correlations of major combustion products observed in the hood experiments depend on the layer temperature as well as the fuel type. This point will be discussed in more detail subsequently.

Morehart tried to address the temperature effect by performing detailed chemical-kinetic calculations of a plug flow reactor for a rich mixture typical of his upper layer. [23] The calculations showed that such mixtures did become reactive for temperatures greater than 700 K in agreement with his findings, but that the calculated changes in upper-layer composition were not consistent with the differences between the Toner and Morehart experiments. Similar calculations performed at NIST will be summarized in section IV.

Morehart also reported limited measurements using ethylene and propylene fuels. Similar to Beyler's results [16],[17],[18], variations in upper-layer concentrations with fuel type were observed. Figures 21 and 22 compare measured mole fractions of CO and $O_2$ as a function of $\phi_g$ for the three fuels. While the general behaviors are quite similar (e.g., CO

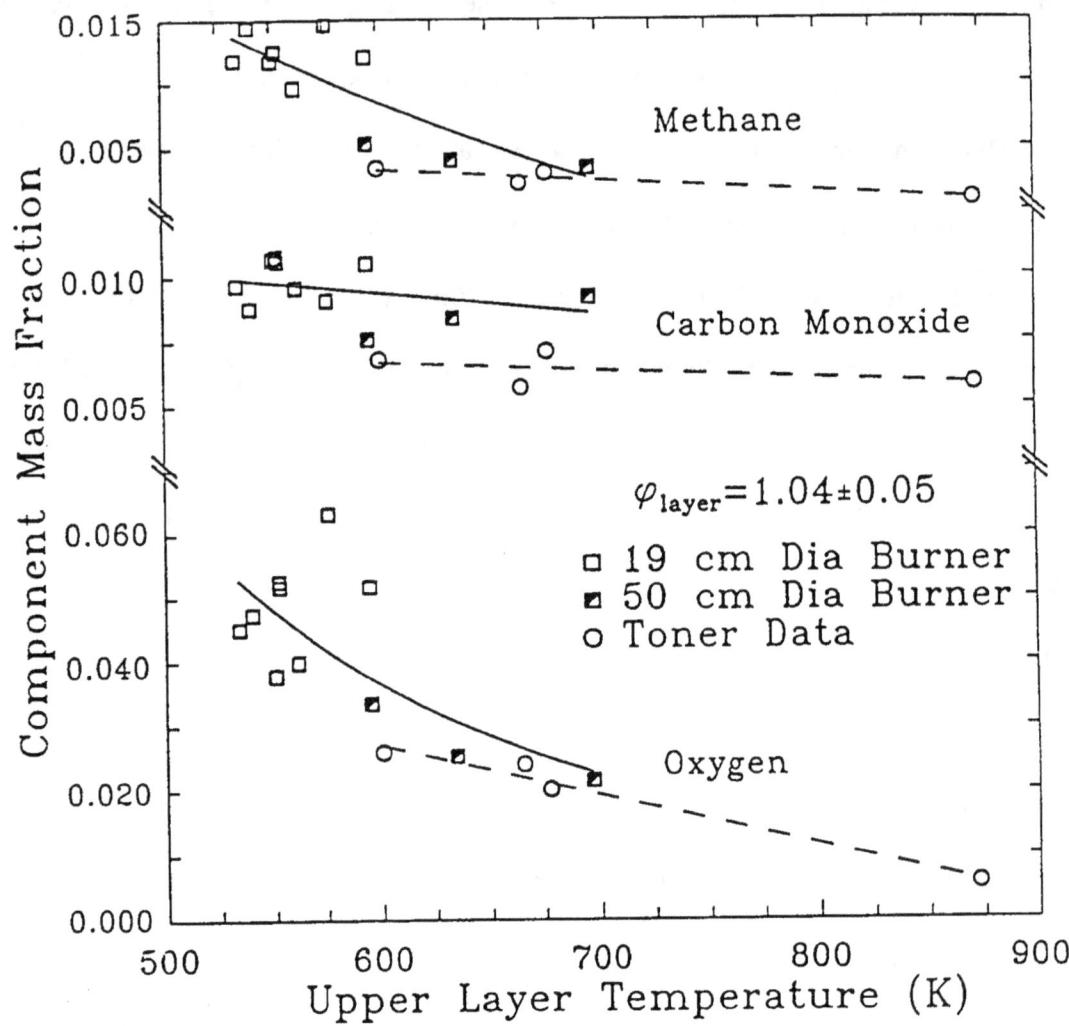

Figure 20. Upper-layer mass fractions of methane, carbon monoxide, and oxygen are plotted as a function of upper-layer temperature for natural gas fires. $\phi_g = 1.45$. Data from the investigation of Morehart [23] are compared with Toner's results [22].

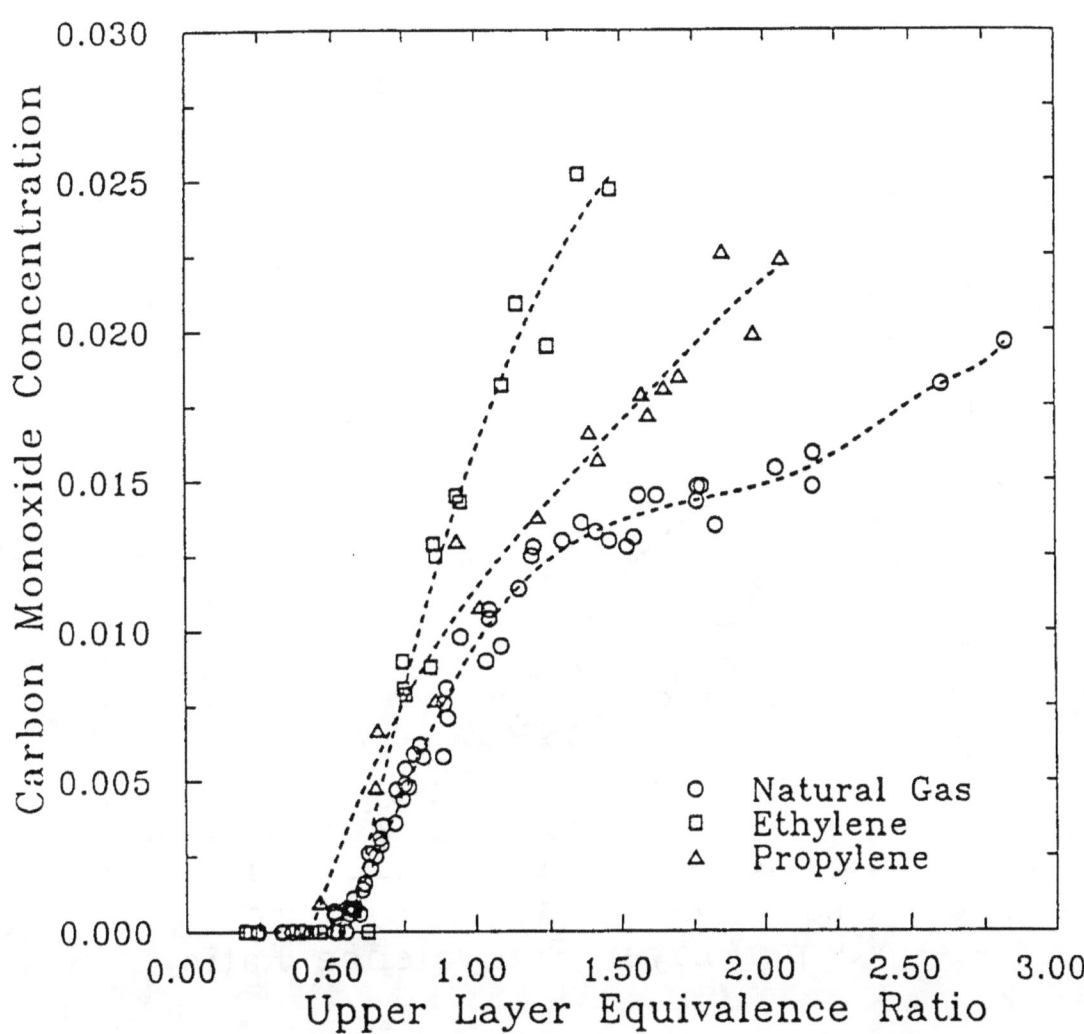

Figure 21. Mole fractions of carbon monoxide observed in the hood experiments of Morehart [23] as functions of $\phi_g$ are shown for natural gas, ethylene, and propylene. Figure taken from [23].

41

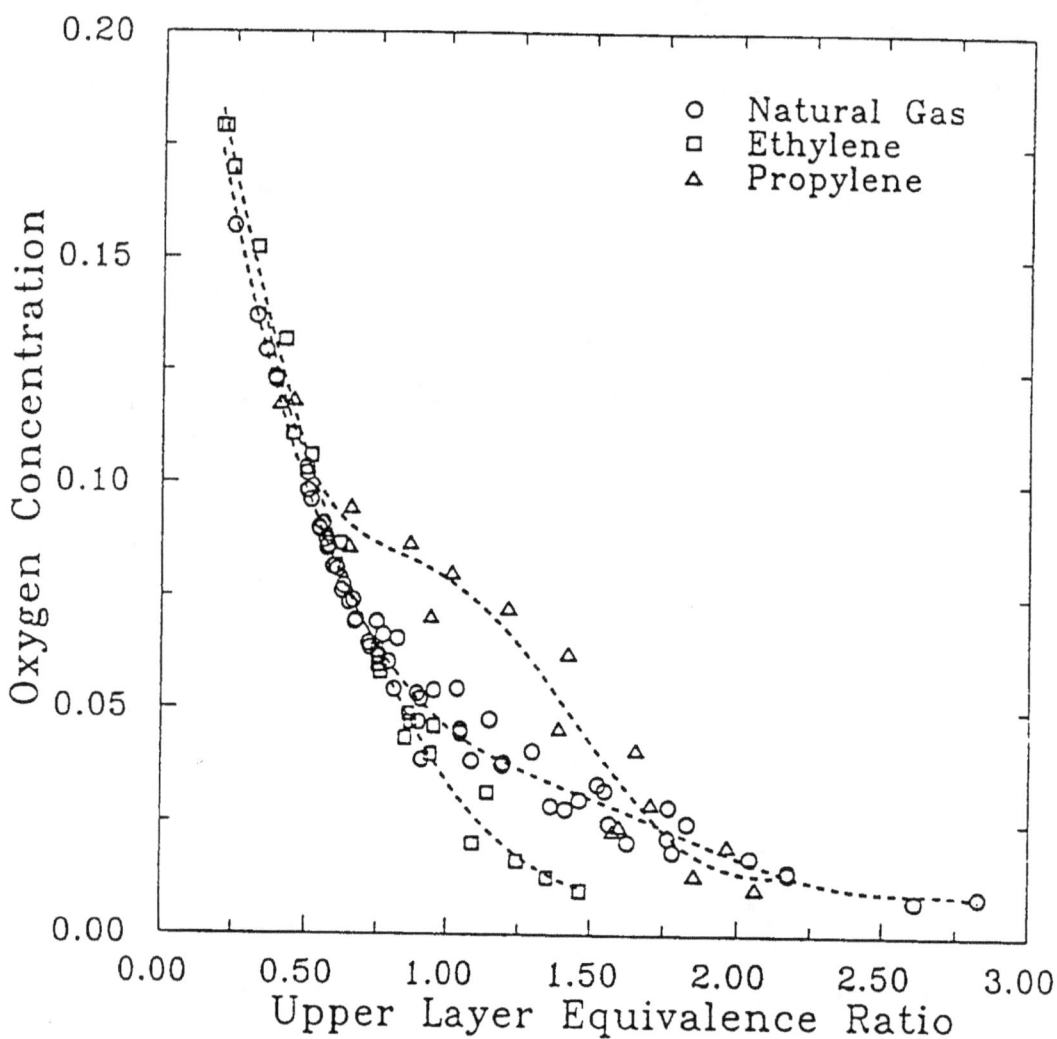

<u>Figure 22.</u>   Mole fractions of oxygen observed in the hood experiments of Morehart [23] as functions of $\phi_g$ are shown for natural gas, ethylene, and propylene. Figure taken from [23].

concentrations begin to rise for $\phi_g > 0.5$, and significant $O_2$ is observed for rich mixtures), large variations are observed in the concentrations of final products.

Beyler also used propylene as a fuel [16],[17]. Morehart compared his results for this fuel directly with Beyler's. Figures 23-26 show these comparisons for $O_2$, $CO_2$, $CO$, and $H_2$ mole fractions as functions of $\phi_g$. There are significant differences between the two sets of measurements, but the variations are very similar to those observed when the Morehart [23] and Toner [22] natural-gas data were compared. For instance, Morehart observes higher oxygen concentrations for $\phi_g > 0.5$, and $CO$ and $H_2$ concentration behaviors are similar in both comparisons even though $CO$ is observed for $\phi_g > 0.5$ while $H_2$ does not start to increase until the upper layer becomes rich. The similarity of the variations for natural gas and propylene suggests that temperature differences are responsible in both cases. Indeed, upper-layer temperatures in the Beyler experiments [16] were at least 200 K higher than in the larger hood of Morehart. The observation of such differences for two experiments in different laboratories greatly strengthens the conclusion of a temperature effect on correlations of combustion gas species as a function of $\phi_g$.

Morehart also addressed the question as to whether chemical equilibrium calculations could be used to predict concentrations observed in the upper layer [23]. Similar to earlier work, he found that equilibrium concentrations did not correlate with the experimentally observed concentrations.

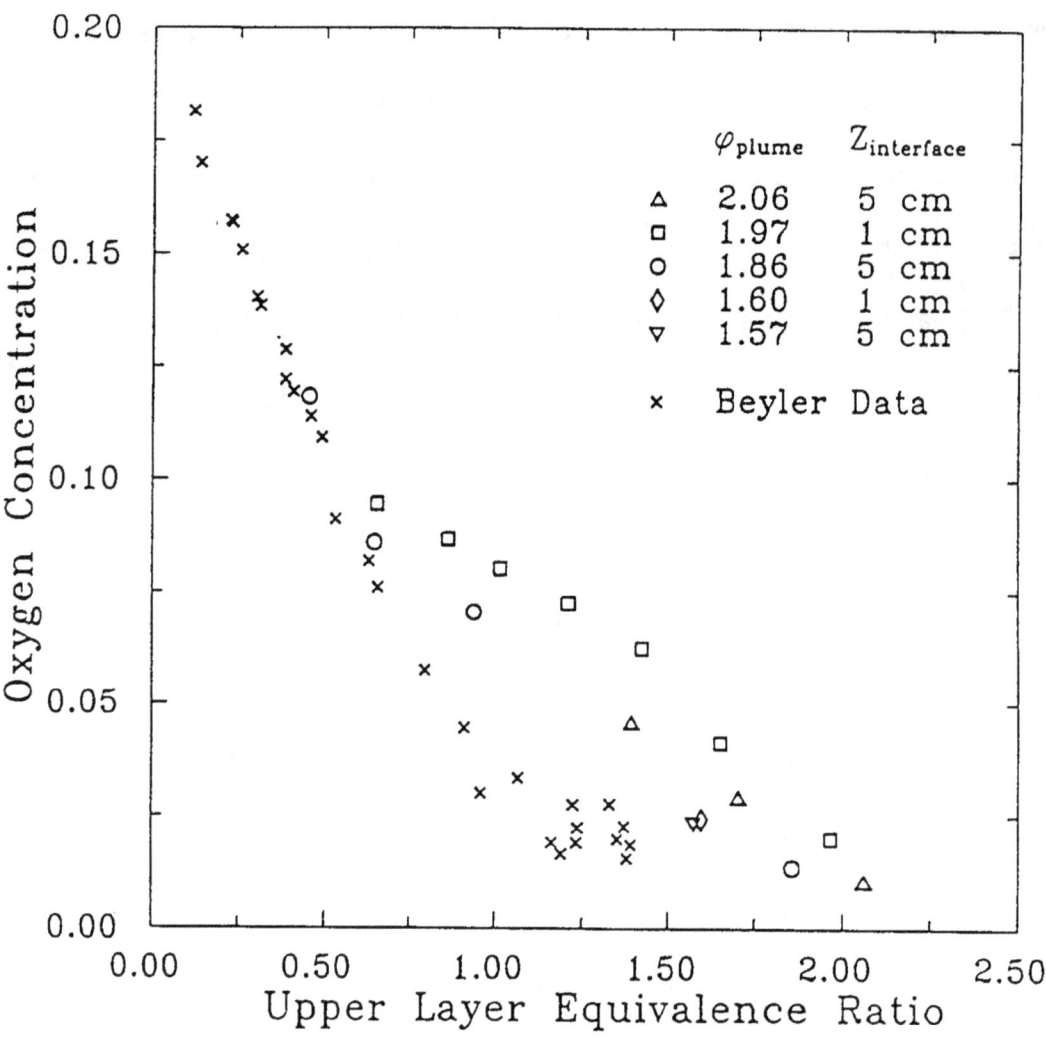

Figure 23.    Mole fractions of oxygen as a function of $\phi_g$ observed in the hood experiments
of Beyler [16] and Morehart [23] are compared.  The fuel was propylene in
both cases.  Figure is taken from [23].

44

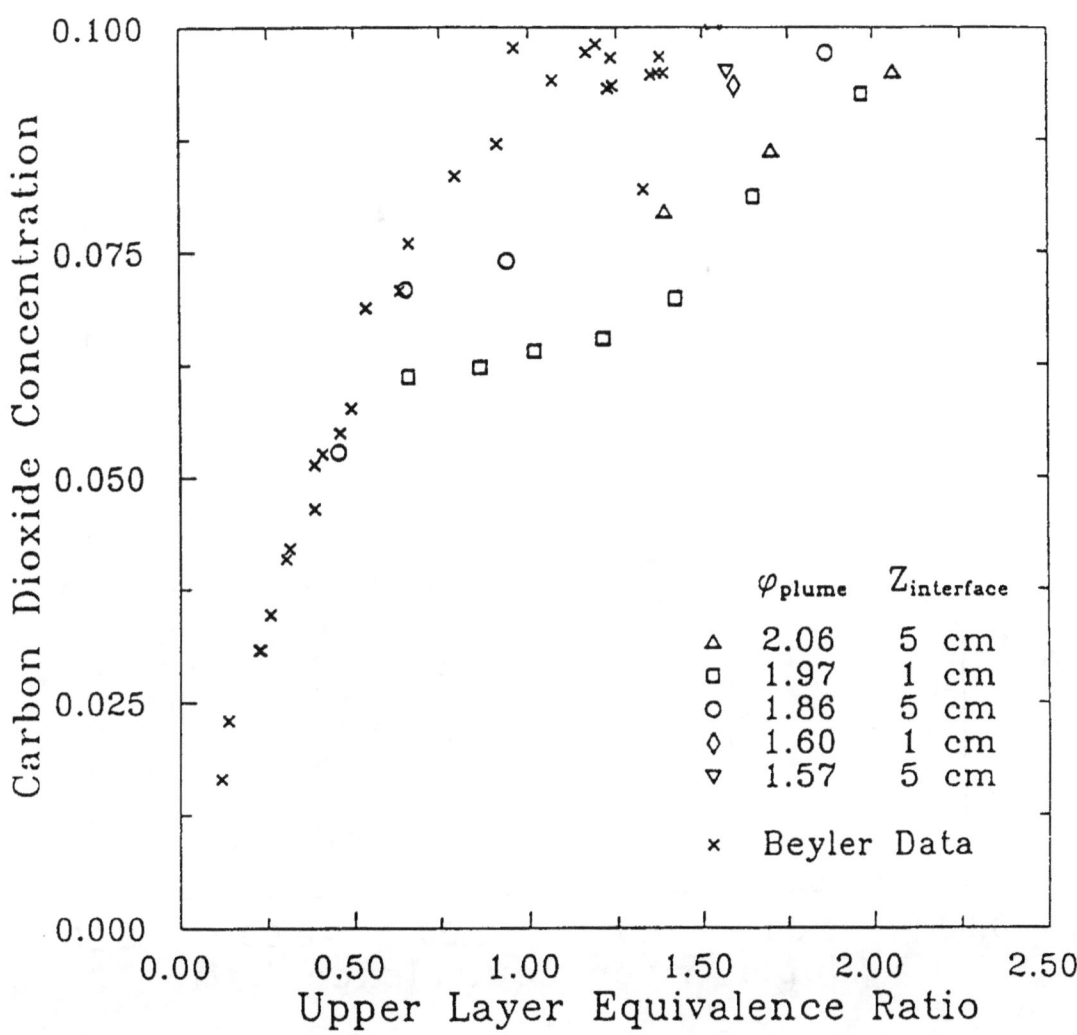

Figure 24. Mole fractions of carbon dioxide as a function of $\phi_g$ observed in the hood experiments of Beyler [16] and Morehart [23] are compared. The fuel was propylene in both cases. Figure is taken from [23].

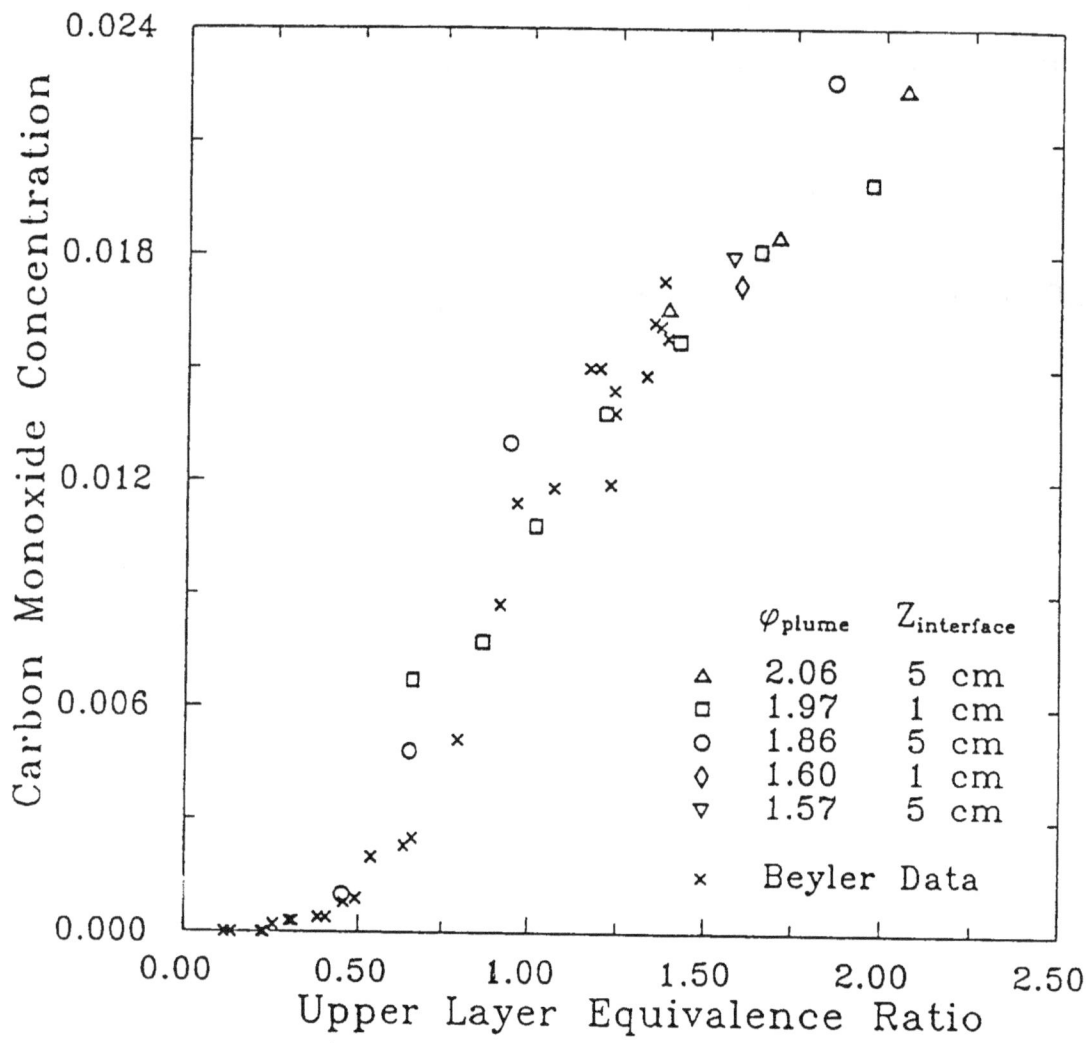

Figure 25. Mole fractions of carbon monoxide as a function of $\phi_g$ observed in the hood experiments of Beyler [16] and Morehart [23] are compared. The fuel was propylene in both cases. Figure is taken from [23].

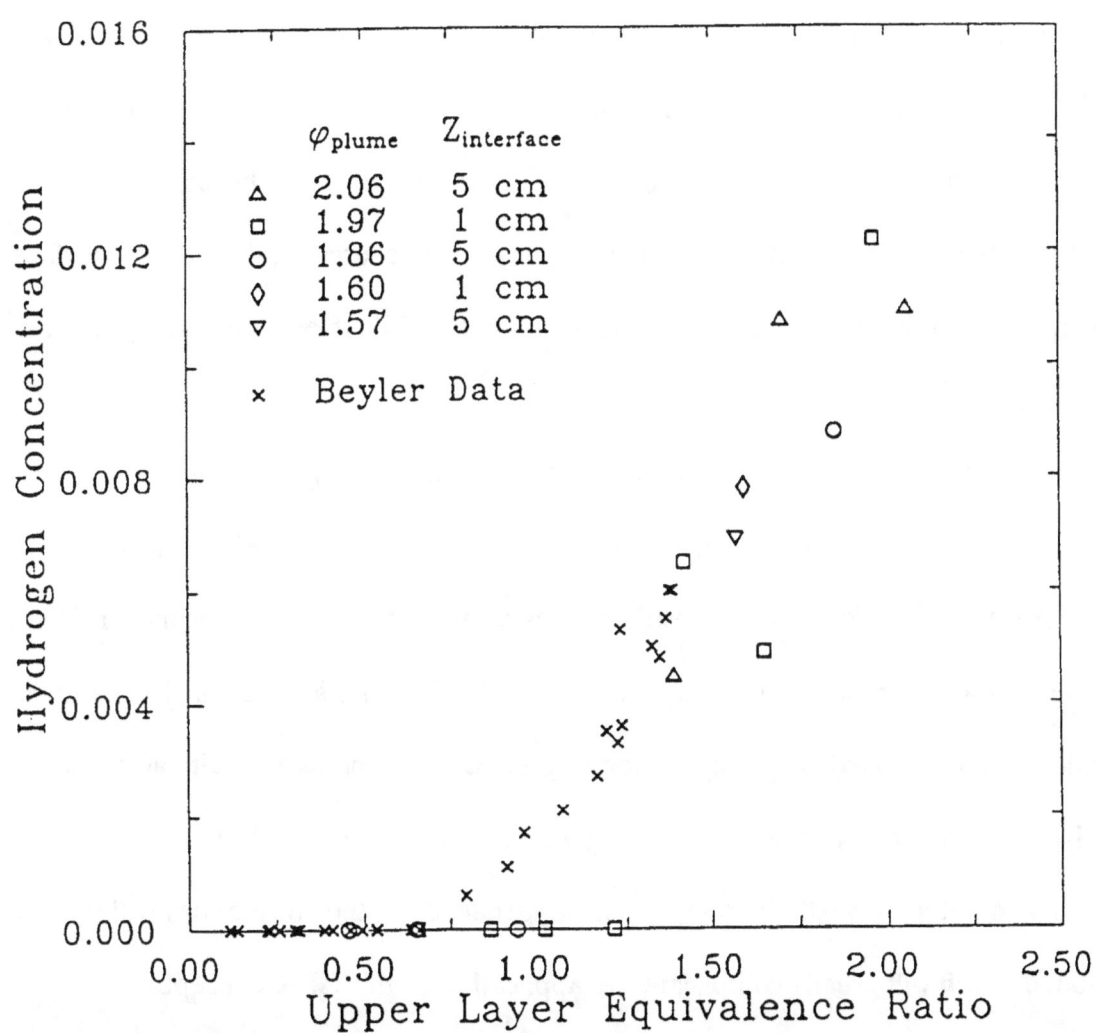

Figure 26. Mole fractions of hydrogen as a function of $\phi_g$ observed in the hood experiments of Beyler [16] and Morehart [23] are compared. The fuel was propylene in both cases. Figure is taken from [23].

## C.    Conclusions Based on the Hood Experiments

The experimental findings discussed above demonstrate that the composition of upper layers in the hoods above simple fires are well correlated by plotting the concentrations in terms of $\phi_g$. This is true even when $\phi_p$ is not the same as $\phi_g$. The correlations are insensitive to fuel supply rate and the separation of the fuel source and the layer interface. However, the correlations are found to depend on the fuel itself and the upper-layer temperature.

The existence of these correlations has been termed the global equivalence ratio (GER) concept. The focus of this manuscript is whether or not the GER concept can be used to predict combustion gas compositions in enclosure fires with a particular emphasis on CO. Before the dependence on temperature was identified, it had been hoped that these correlations could be used simply by performing hood experiments for suitable fuels and using the resulting correlations to predict generation rates for the fires. Even if this approach is found to be valid, the temperature dependence of the correlations will need to be included which will greatly complicate the application of the GER concept.

Even though the data are limited, there does seem to be some consistency in the dependence of the GER correlations on temperature. For upper-layer temperatures lower than 500 K, reactions seem to be unimportant and the correlations are independent of temperature. As the temperature is increased in the range 500-700 K, shifts in the composition of the products are observed. The data suggest that the changes in composition are the result of oxidation of additional fuel to produce CO, $CO_2$ and $H_2O$. Concentrations of $H_2$ seem to be relatively insensitive to the temperature effect.

The degree to which oxygen reacts appears to be well correlated with temperature. Figure 23 shows that Beyler and Morehart both measured significant levels of $O_2$ for rich upper layers, but that the concentrations observed by Beyler were lower as expected based on the higher temperatures present in the Beyler hood. The decrease in $O_2$ concentration extends down to $\phi_g = 0.5$ where the first products of incomplete combustion are observed. On the other hand, the Toner measurements (which had the highest upper-layer temperatures (maximum of 870 K)) show that the layer is nearly depleted of oxygen for rich conditions.

The dependence of the upper-layer concentrations of CO and $CO_2$ on temperature is not as clear. Figures 15 and 24 indicate that at higher temperatures the concentrations of $CO_2$ increased at rich conditions for both Beyler's and Toner's data as compared to the lower temperature results of Morehart. Strangely, the CO measurements at rich conditions for Beyler and Morehart (fig. 25) are in good agreement while Toner measured higher concentrations of CO for rich conditions (fig. 16). These observations suggest that shifts in the relative compositions of these partially oxidized species are to be expected with increasing temperature. A similar dependence on temperature is found for comparisons of CO concentrations under lean upper-layer conditions. The data of Beyler and Morehart are in good agreement while Toner's results indicate that much lower CO levels are present.

It should be noted that the focus of the source of changes in the upper-layer composition has been temperature. However, since the modification in concentration

49

appears to vary slowly with temperature it might also be expected that the average residence time in the upper layer may also be an important variable.

Considerably more data will be required to fully characterize the temperature dependence of upper-layer compositions in the hood experiments. The temperature dependence is clearly a difficulty in applying the GER concept to enclosure fires. However, the correlations which are observed are quite robust. The limited hood data available seem to suggest that as upper-layer temperatures approach 900 K the $O_2$ concentrations in the upper layers will approach zero. If this is the case, it is possible that additional reactions will not occur, and that unique correlations of upper-layer composition with $\phi_g$ may exist for the rich conditions which are characteristic of the most dangerous enclosure fires. Since upper-layer temperatures in fully developed enclosure fires tend to be higher than 900 K, it is concluded, based **solely** on the results of the hood experiments, that the GER concept may be applicable to enclosure fires if appropriate corrections are incorporated for temperature effects.

## IV. DETAILED CHEMICAL-KINETIC CALCULATIONS OF UPPER-LAYER REACTION BEHAVIOR

### A. Summary and Discussion of Calculational Results

A series of detailed chemical-kinetic calculations has been performed to characterize the expected reaction behavior of the upper-layer gases observed in Morehart's natural-gas hood experiments [23],[24],[25] when introduced into reactors having much higher temperatures characteristic of upper layers in fully developed room fires. A summary of the calculational approach and results is available [29]. This study and its implications for application of the GER concept are summarized here. As noted earlier, a limited study along the same lines was done by Morehart [23].

The rationale for the investigation was that understanding the reaction behavior of gas compositions typically found in upper layers would aid in the assessment of the possibility of using the GER concept to predict upper-layer composition in enclosure fires. In order for the correlations derived from the hood experiments to be valid in the higher temperature environments (up to 1300 K) typical of upper layers in enclosure fires, two minimum requirements must be fulfilled:

1. The relative generation rates of combustion products by the fire plume entering the upper layer of an enclosure fire must be unaltered from those characteristic of the hood experiments.

and

2. The gases generated by the fire plume must be nonreactive at the higher temperatures typical of enclosure fires.

The second of these requirements is tested by the calculations discussed here. In the previous section, it was noted that the hood experiments had shown that changes in the final product distribution of the combustion gases are observed as the hood temperature is increased. The calculations provide insight as to whether the reactions responsible are taking place in the fire plume or in the gas mixtures of the upper layer located outside of the plume.

Calculations were performed over a temperature range of 700-1300 K. The values of $\phi_g$ chosen for the calculations were those for which products of incomplete combustion were observed (0.5-2.83) in the hood experiments. Since the heat loss and mixing behaviors of upper layers are quite complicated and difficult to model, idealized cases corresponding to limits for these parameters were used. Calculations were carried out assuming two conditions: (a) instantaneous heat transfer to or from the walls of the enclosure such that the temperature of the layer remains constant (isothermal case) and (b) no heat transfer occurs to the enclosure (adiabatic case). Mixing within the upper layer was assumed to be either infinitely fast (perfectly stirred reactor (PSR) model) or infinitely slow (plug-flow reactor model).

Computer codes provided by the Combustion Research Facility of Sandia National Laboratory were used for the calculations [30],[31],[32]. Following a careful review of the literature, a chemical reaction mechanism formulated by Dagaut et al. [33] for the oxidation of ethylene was chosen for use in the modeling. This mechanism was chosen because it included one- and two-carbon species, it had been recently updated with the latest available rate constants, and was validated by comparison with jet-stirred reactor measurements at temperatures (880-1253 K) and equivalence ratios (0.15-4) similar to those of interest here. The mechanism includes 31 chemical species undergoing 181 chemical reactions. No species with more than two carbon atoms are incorporated and heterogeneous chemistry is assumed to be unimportant. The latter assumption should be good for natural-gas fires which generate low levels of soot. Thermodynamic values required for the calculations were taken from tables provided by the Combustion Research Facility [34].

For the calculations, initial concentrations of gases for the reactors were taken directly from tabulations of combustion products observed in the upper layers of natural-gas fires

Table 2. Calculational Matrix for Detailed Chemical-Kinetic Calculations

| PROPERTY | CONDITIONS TREATED |
|---|---|
| Molecular Mixing | Infinitely Slow (Plug-Flow Reactor) and Infinitely Fast (Perfectly Stirred Reactor) |
| Heat Transfer Behavior | Isothermal and Adiabatic |
| Temperature | 700 - 1300 K in steps of 100 K |
| Equivalence Ratios | $\phi_g$ = 0.50, 0.81, 1.09, 1.30, 1.62, 1.76, 2.04, 2.17, 2.61, and 2.83 |
| Residence Time | 0 - 20 s |

by Morehart [23]. The time behavior of the reactors for a chosen initial temperature was then calculated for residence times covering a range of 0-20 s. Table 2 summarizes the ranges of conditions over which calculations were performed.

Clearly, the calculations are too extensive to discuss in full detail. Results will be shown which are representative of the calculated behaviors, and the major trends will be summarized. Figure 27 shows plots of calculated mole fractions versus time for a plug flow reactor and isothermal conditions. Results for $\phi_g$ = 2.17 and reactor temperatures of 800 K and 1300 K are shown. Note the different time scales for the two cases. The reactions at 800 K occur over a period greater than 20 s, while at 1300 K the reactions are nearly complete in less than one second.

There are significant differences in the reaction behavior for the two temperatures. In both cases the $CO_2$ levels remain nearly constant while CO increases. However, less CO is generated at the lower temperature. Similarly, the $H_2$ concentration barely changes at 800 K while it is significantly increased at 1300 K. Detailed comparisons of the results show

53

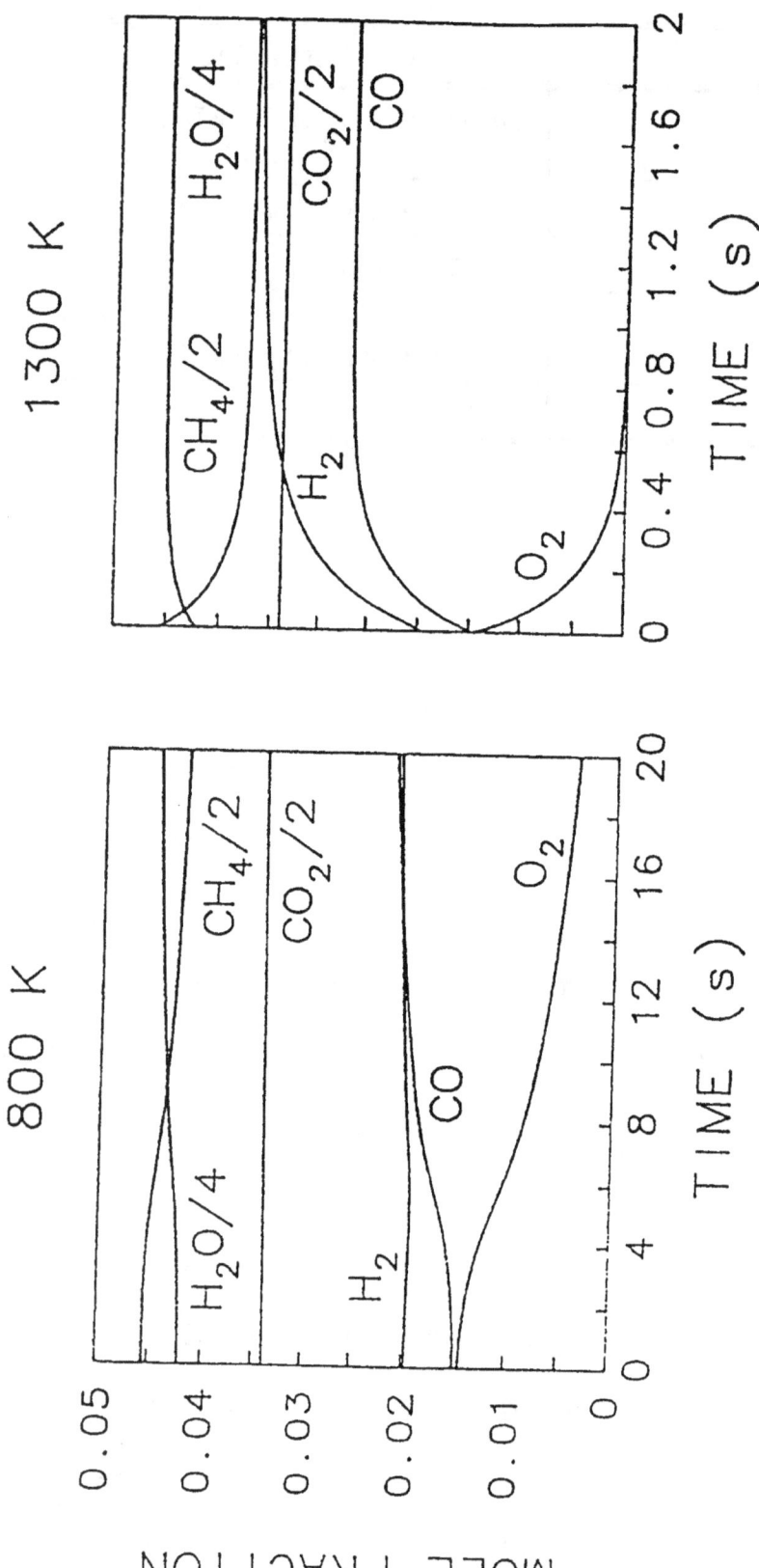

Figure 27.    Calculated time behavior of major gas species mole fractions for an isothermal plug-flow reactor as a function of residence time for $\phi_g$ = 2.17.   Results for 800 K and 1300 K are shown.  Note the short time scale for the higher temperature.   Initial mole fractions taken from [23] are: nitrogen (0.62), oxygen (0.014), water (0.17), carbon dioxide (0.068), methane (0.091), carbon monoxide (0.015), hydrogen (0.020), ethane (0.0014), and acetylene (0.0017).

that for the high temperature case more oxygen is available for reaction with fuel molecules due to the formation of hydrogen instead of water. The oxidation of the organic fuels generates primarily additional CO, and there is very little oxidation of CO to $CO_2$ for these rich conditions.

Figure 28 shows the time behavior of CO mole fractions calculated over a range of temperatures for $\phi_g = 2.17$. The major conclusion from this figure is that the final concentrations of CO fall into two well defined regimes. For temperatures less than 1000 K the increase in CO is $\approx 40\%$ while for temperatures greater than 1100 K the increase is $\approx 80\%$. There is a clear transition between 1000 K and 1100 K. This transition is caused by the shift in the distribution of products between $H_2$ and $H_2O$ as discussed above.

Figure 29 shows the time behavior of major products for $\phi_g = 1.09$ and with the other parameters the same as above. At 800 K the reaction is very slow, and significant oxygen and fuel remain after 20 s. For this nearly stoichiometric mixture both CO and $CO_2$ are produced while $H_2$ is little changed. At 1300 K the behavior is very different. Initially, methane is oxidized to generate primarily CO and $H_2$ as observed for the high temperature rich case. Ultimately the methane and other fuel molecules are depleted, and there is a sudden, rapid reaction of the $H_2$ and CO with the remaining $O_2$ to generate $H_2O$ and $CO_2$. This result clearly shows that as long as fuel is present the oxidation of CO is strongly inhibited.

The calculated time dependencies of CO concentration for a range of $\phi_g$ are shown in figure 30 for an isothermal plug-flow reactor at 1000 K. For lean conditions, CO is formed as long as excess fuel is present, but then rapidly reacts to form $CO_2$ when the fuel

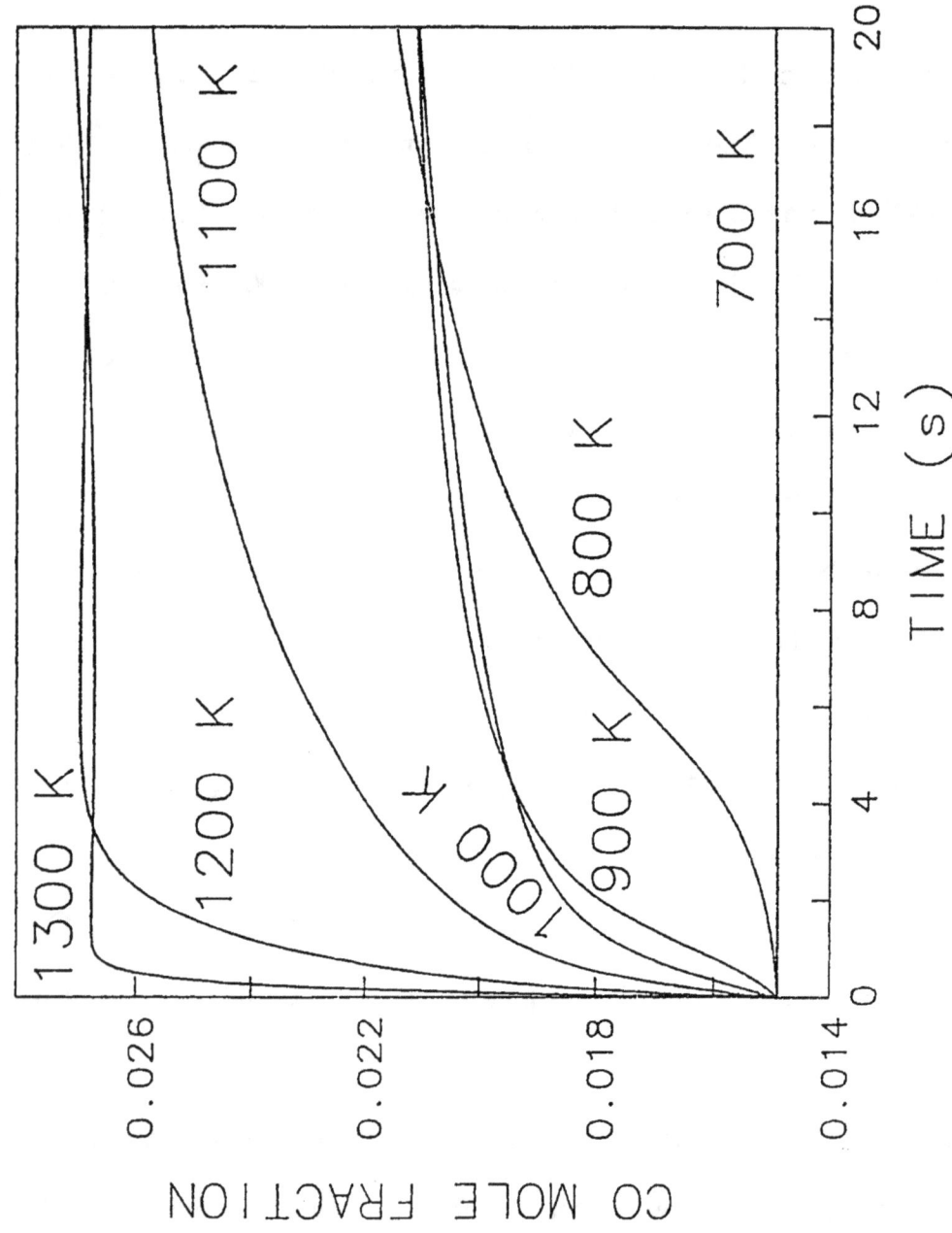

Figure 28.    Calculated carbon monoxide mole fractions as a function of time for an isothermal plug-flow reactor at the temperatures indicated. $\phi_g$ = 2.17. Initial concentrations are the same as in figure 27.

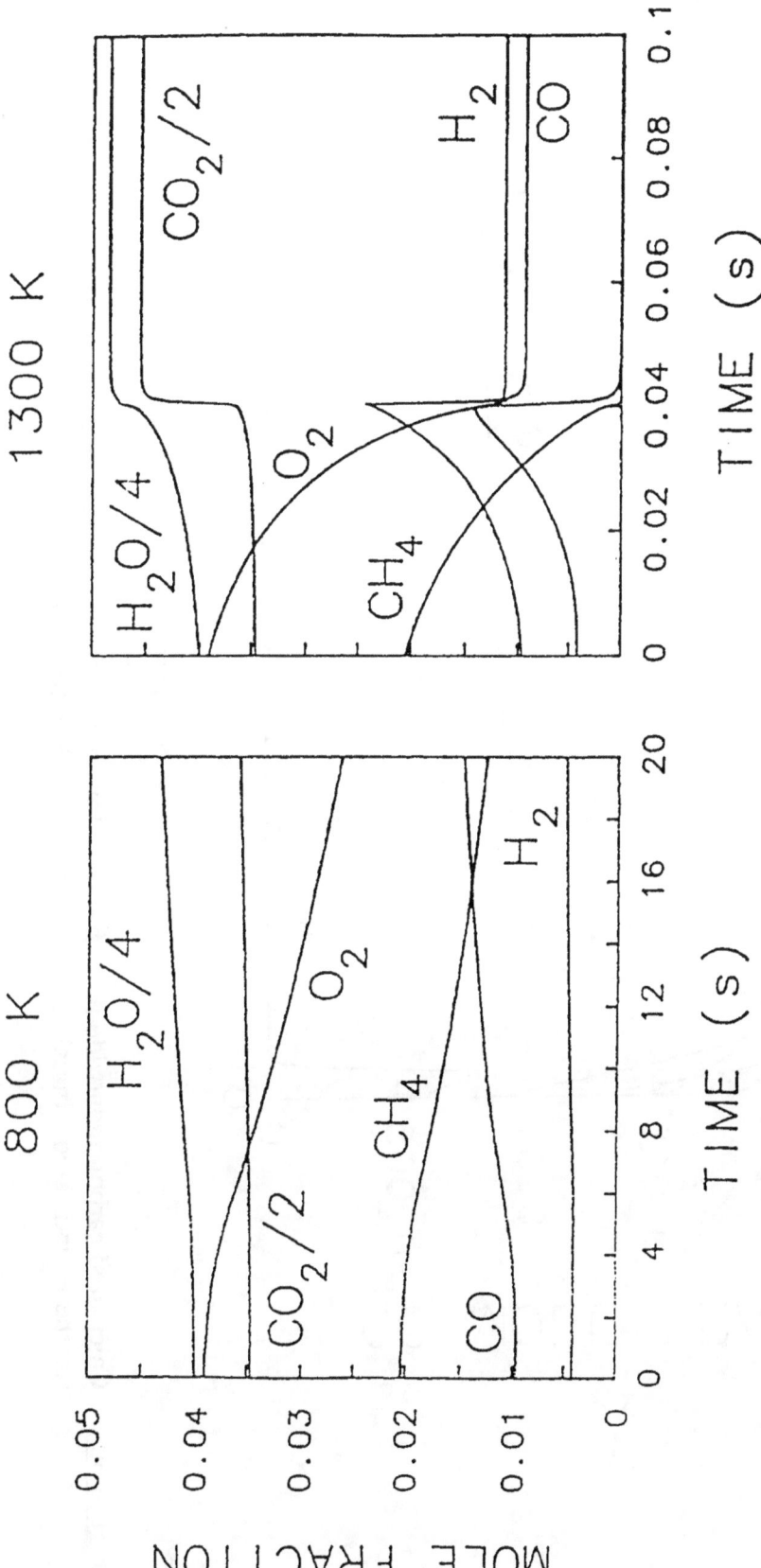

Figure 29. Calculated time behavior of major gas species mole fractions for an isothermal plug-flow reactor as a function of residence time for $\phi_g = 1.09$. Results for 800 K and 1300 K are shown. Note the short time scale for the higher temperature. Initial mole fractions taken from [23] are: nitrogen (0.69), oxygen (0.69), water (0.16), carbon dioxide (0.069), methane (0.020), carbon monoxide (0.010), hydrogen (0.004), ethane (0.0000), and acetylene (0.0007).

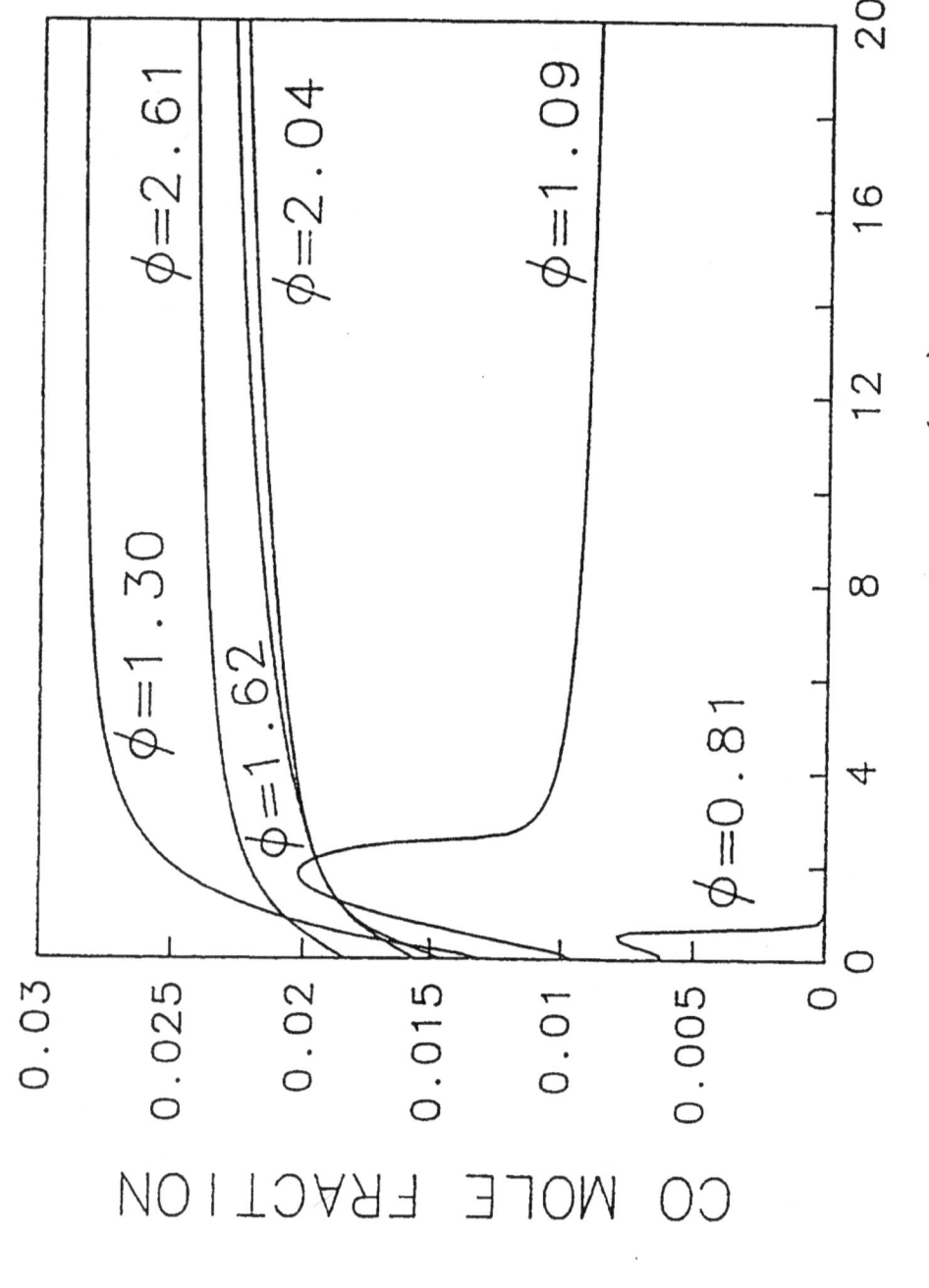

58

<u>Figure 30.</u>　Calculated carbon monoxide mole fractions for an isothermal plug-flow reactor at 1000 K as a function of time for the $\phi_g$ indicated. Initial concentrations are taken from Morehart [23].

is depleted. The oxidation of CO then continues until the oxygen is depleted. For $\phi_g$ which are just rich, this competition results in significant concentrations of CO remaining following reaction. For still higher $\phi_g$, the oxygen is depleted before any significant conversion of CO to $CO_2$ takes place with the result that the CO concentration is increased significantly by oxidation of fuel. The largest fractional increase in CO concentration is for $\phi_g = 1.30$, since this case provides the most $O_2$ for converting fuel to CO without depleting the excess fuel.

Comparisons of calculated behaviors for the PSR and plug flow reactors for identical starting conditions showed only minor differences, indicating that the reaction behavior is weakly dependent on the mixing behavior. Figure 31 shows such comparisons for $\phi_g = 1.76$ and temperatures of 900 K and 1200 K.

The dependence on the heat loss behavior was more complicated. Figures 32 and 33 compare calculations of the CO mole fraction as functions of time for a plug flow reactor having $\phi_g = 1.76$ and assuming both isothermal and adiabatic conditions. For initial temperatures of 900 K and 1200 K (fig. 32) the results for the two cases are quite comparable, even though there is a significant increase in temperature for the adiabatic cases, since the oxidation of the fuel to form CO releases heat. On the other hand, the ultimate concentrations of CO are much higher for the adiabatic case when the initial starting temperatures are 1000 K and 1300 K (fig. 33).

The explanation for increased CO in the 1000 K case is straightforward. The heat release raises the temperature to the point where the transition in reaction behavior, observed between 1000 K and 1100 K for the isothermal cases, takes place, and the formation of $H_2$ becomes more favorable than $H_2O$. Note that the final CO concentration

59

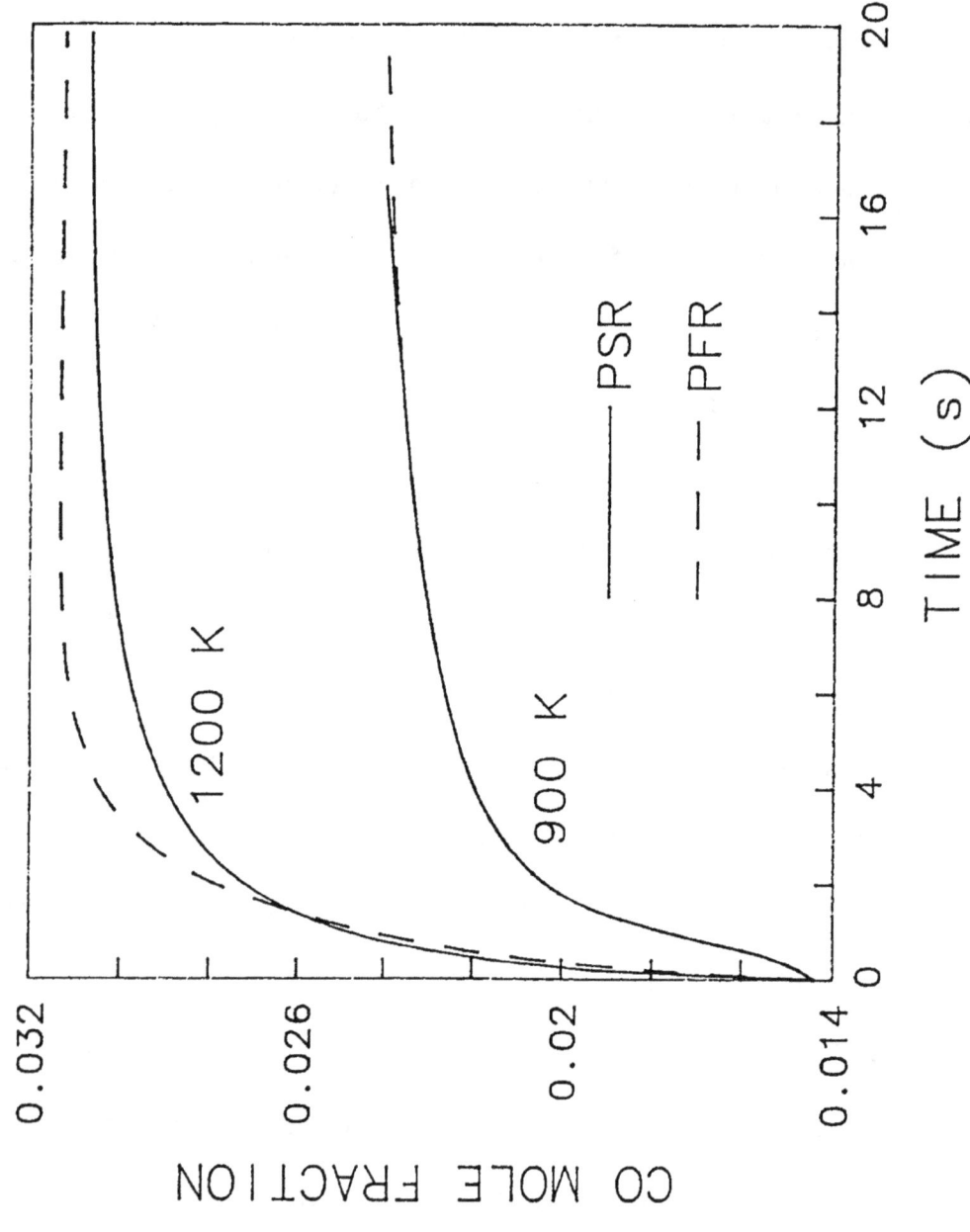

Figure 31.   Comparison of calculated carbon monoxide mole fraction as a function of time for a perfectly-stirred reactor and a plug-flow reactor having initial temperatures of 900 K and 1200 K. $\phi_g$ = 1.76.  Initial concentrations taken from Morehart [23].

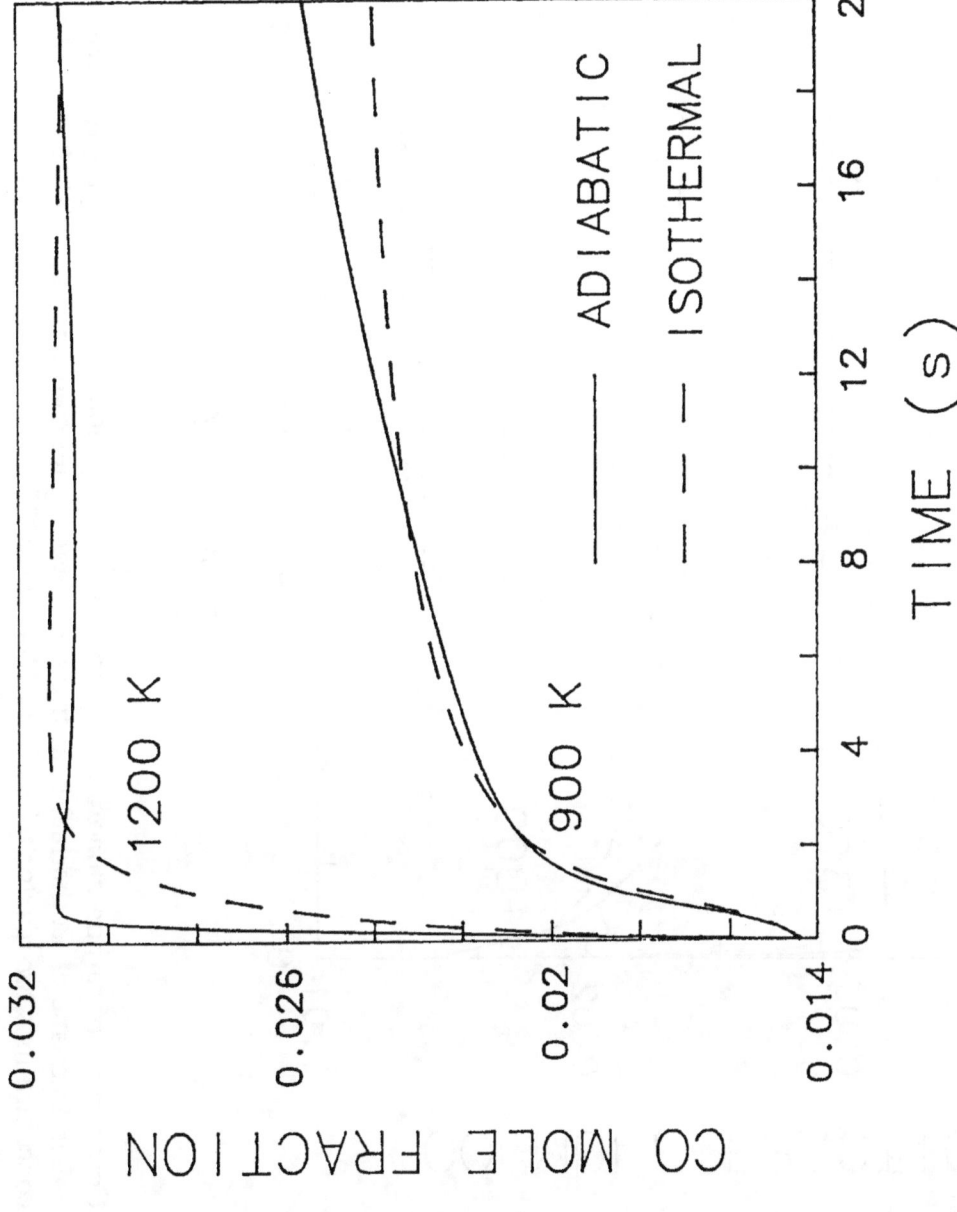

Figure 32. Comparison of carbon monoxide mole fraction as a function of time for a plug-flow reactor calculated assuming either adiabatic or isothermal heat transfer. Starting temperatures are 900 K and 1200 K. $\phi_g = 1.76$. Initial concentrations for reactor taken from Morehart [23].

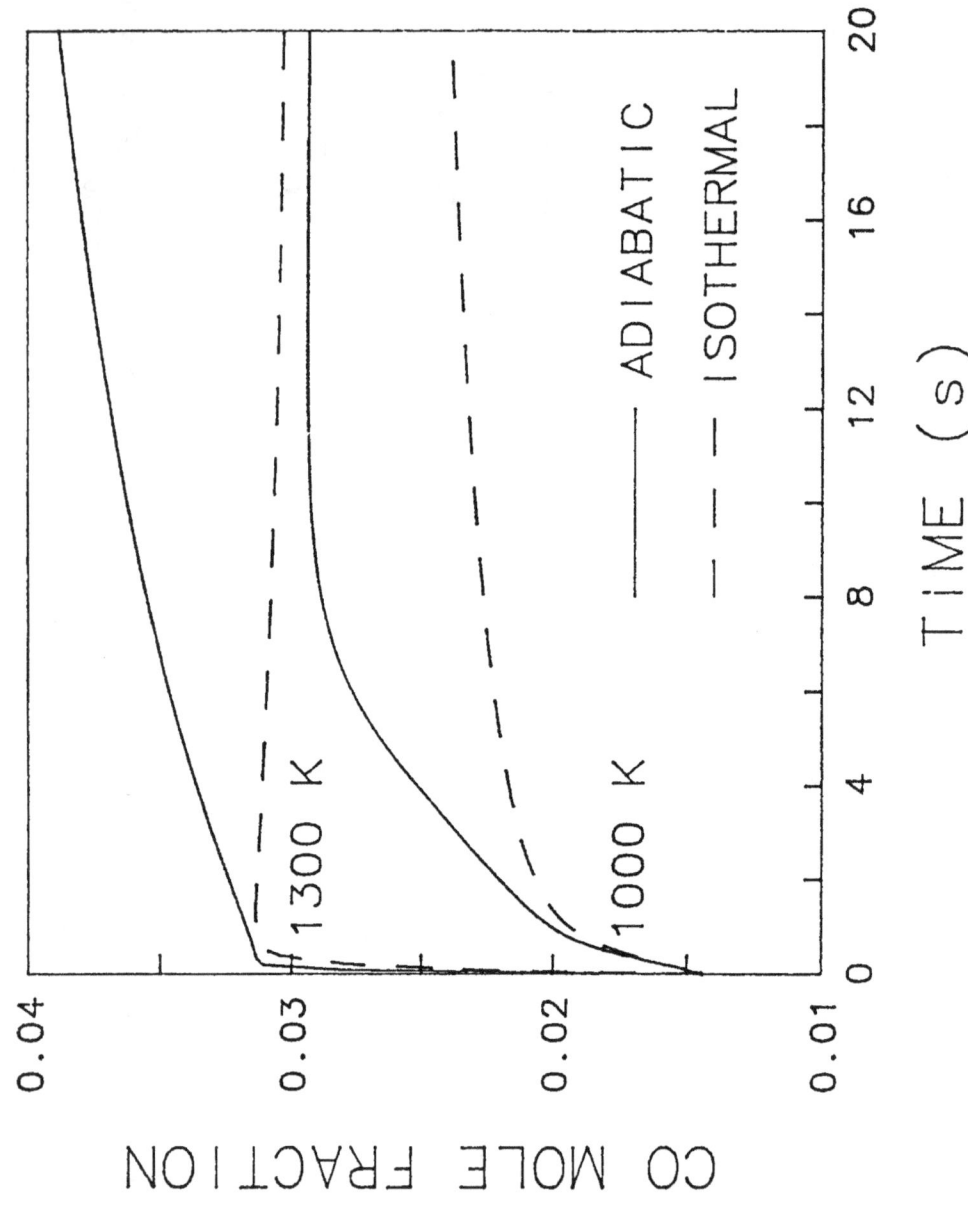

Figure 33. Comparison of carbon monoxide mole fraction as a function of time for a plug-flow reactor calculated assuming either adiabatic or isothermal heat transfer. Starting temperatures are 1000 K and 1300 K. $\phi_g$ = 1.76. Initial concentrations for reactor taken from Morehart [23].

for the adiabatic calculation with an initial temperature of 1000 K is very similar to the isothermal case at 1300 K. The reason for the increase in CO concentration for the adiabatic calculation with an initial temperature of 1300 K is more subtle. As the initial adiabatic reaction proceeds, the reaction mixture temperature rises above 1300 K. Examination of the detailed reaction behavior shows that for temperatures significantly above 1300 K, the water gas shift reaction,

$$CO + H_2O = CO_2 + H_2$$

starts to come into equilibrium slowly. At these high temperatures the formation of CO is strongly thermodynamically favored; and $CO_2$ and $H_2$ are converted to CO and $H_2O$, resulting in the observed increase in CO while simultaneously cooling the reaction mixture.

### B.    Implications of the Calculations for the Results of Hood Experiments

The calculations summarized here provide important insights into the experimental observations for the hood experiments. It was pointed out above that raising the temperature of hood gases led to a depletion of oxygen and increased concentrations of CO and $CO_2$ even though the relative changes in CO and $CO_2$ were not well defined. The most important observation is that concentrations of $CO_2$ are significantly increased as the hood temperature becomes higher. This is contrary to the expected behavior if the gas mixtures observed in the low-temperature hood experiments are simply introduced into a higher temperature environment. The detailed chemical-kinetic calculations clearly indicate that the formation of CO would be highly favored, and the concentration of $CO_2$ would be expected to remain nearly constant.

It must be concluded that the higher hood temperatures lead to a modification in the generation rates of chemical species by the fire plume, with the result that the fuel is more fully oxidized than when temperatures in the hood are very low. Since it is currently impossible to model the formation of chemical species in a quenched fire plume, this result suggests that it will not be possible to model the variations of GER concentration profiles with temperature. Experimentation will be required to characterize the effect.

It was noted earlier that temperatures in excess of 1300 K are required for the water gas shift reaction in the hood gases to start to come into equilibrium. The need for such high temperatures to equilibrate these chemical species explains why attempts to predict concentrations of chemical species in the hood experiments using chemical equilibrium arguments failed. The necessary reactions freeze out at temperatures much higher than the kinetically controlled reactions discussed up to this point. The final product distribution must be expected to be far from equilibrium.

C.    **Implications of the Calculations for CO Formation in Upper Layers of Enclosure Fires**

The detailed chemical-kinetic calculations yield the expected reaction behavior if oxygen is somehow mixed into a rich upper layer formed by a fire. Since these layers typically have temperatures greater than 1100 K and are regions of very rapid mixing, CO and $H_2$ will be formed very efficiently. From a fire toxicity standpoint, this is the worst possible scenario since it suggests that if such mixing takes place, significantly higher concentrations of CO will be present than would result simply from quenching of the fire plume reactions upon entering the upper layer.

Nakaya has proposed a model for the formation of CO within the upper layers of enclosure fires based on the assumption that combustion gases in the upper layer are in chemical equilibrium [35]. The detailed chemical-kinetic calculations have shown that temperatures well in excess of 1300 K are required in order for the major combustion products to come into equilibrium. The highest temperatures generally observed in enclosure fires are on the order of 1300 K, suggesting that the assumption of full chemical equilibrium will be inappropriate; except, possibly, in the case of extremely hot fires. Such hot fires would be expected to generate quite high levels of CO indeed.

## V.  STUDIES IN REDUCED-SCALE ENCLOSURES

### A.  Historical Results Available in the Literature

As noted in Section II there are no systematic large(full)-scale investigations which have considered the formation of CO in enclosure fires. There is, however, a body of literature dealing with CO in reduced-scale enclosures which will be discussed in this section. Additional work which is ongoing at VPISU and at NIST will also be summarized in later subsections.

65

A very early attempt to investigate the production of CO and other combustion gases by underventilated burning within an enclosure was the work of Rasbash and Stark [36]. These researchers burned cellulosic fuels within a cube having 0.91 m sides. Most experiments were performed using wood wool suspended in spherical cages of various diameters at the center of the cube. Ventilation was provided by slits of variable width located along the top edge of one wall. Note that this configuration would be expected to hinder the development of a well-defined two-layer system.

Measurements of fuel species at various locations within the cube gave the same results, except for locations directly above the fuel source. In most experiments temperatures were quite low, becoming greater than 700 K for only a few cases. Dry concentration data were well correlated in terms of the parameter $Ah^{1/2}/W$ where A is the area of the ventilation opening, h is the height of the ventilation opening, and W is the fuel load. Concentrations as high as 10% CO were observed for small openings and large W and decreased to near zero for large openings and small fuel loadings. Increases in CO were observed when the fire became underventilated. Note that the maximum CO concentrations observed are many times larger than those observed by Beyler in his hood experiments burning wood [18].

Oxygen concentrations were also found to be dependent on $Ah^{1/2}/W$, going from near 21% at large values of this parameter down to values approaching 1% for small values. In the vast majority of experiments mole fractions of $O_2$ were greater than 5%.

Gross and Robertson made experimental measurements in three geometrically similar enclosures (width:length:height = 1:2:1) having short sides of 0.16 m, 0.47 m, and

1.5 m [37],[38]. Various sizes of wood cribs were used as the fuel. Ventilation was provided by horizontal or vertical windows located in the center of a side and extending across the entire wall. Data were correlated in terms of the normalized ventilation parameter $Ah^{1/2}/S^2$, where S is the physical-size scaling ratio for the different enclosures.

Gas temperature measurements were found to be relatively uniform in the enclosures except in lower regions where the air entered the vent. Time-averaged temperatures in the enclosures were very low for small ventilation parameters and increased linearly with ventilation parameter until plateaus were reached for values of $Ah^{1/2}/S^2 > \approx 3000$ $cm^{5/2}$. The plateau temperatures had a strong dependence on scale going from $\approx$ 500 K to 1000 K with increasing enclosure size.

Dry concentration measurements for $CO_2$, CO, and $O_2$ as a function of $Ah^{1/2}/S^2$ were reported for the two largest enclosures. Oxygen was found to decrease very rapidly with normalized ventilation parameter and to reach values very close to zero for $Ah^{1/2}/S^2 > 1000$ $cm^{5/2}$. Carbon dioxide concentration showed a similar dependence with an asymptote of $\approx$ 19% for the midsized enclosure and 17% for the largest enclosure. CO concentration behavior was also different for the two enclosures. In the midsized enclosure, CO levels were found to increase with ventilation parameter, reaching maximum values > 12%. In the largest enclosure, the CO concentration was only mildly dependent on ventilation parameter and attained maximum values of 8%.

The measurements were made by extracting gases through an uncooled copper tube. When Gross and Robertson compared their results with measurements made by extracting upper-layer gases through a water-cooled probe, they measured significantly lower concentrations of both CO and $CO_2$ suggesting that additional chemical reactions took place within the uncooled copper probe [37],[38]. Interestingly, in the cooled probe, the formation of CO was favored compared to the uncooled case. Rasbash and Stark have argued that copper and iron tubes may provide a considerably more reactive environment than stainless steel tubes, suggesting that additional probe reactions might be important when copper or iron are used [36]. In an experimental test, Morehart showed that no reactions occurred in a stainless-steel sampling tube for a simulated mixture of upper-layer gases heated to his hood temperature, while significant oxidation of CO to $CO_2$ was observed when a copper tube was used [23]. He concluded that the copper tubing promoted the oxidation reaction, and that the effect was most likely due to catalytic effects on the copper surface. Beyler has hypothesized that Gross and Robertson extracted gases from a reacting zone within the enclosure, and that the gases continued to react in the probe [18]. Morehart's findings suggest this conclusion should be considered suspect since reactions were likely to occur on the metal surface within the probe. It is therefore not necessary to assume reacting gases were sampled.

Tewarson investigated the formation of combustion gases in two geometrically similar enclosures having 1:2:1 aspect ratios with sides of 0.5 m and 1 m, respectively [39]. Ventilation was provided by dual, full-width windows located in the walls on either end of the long dimension of the enclosure. Wood cribs were suspended from the ceilings of the enclosures and were weighed during combustion. The degree of ventilation was varied by changing the heights of the windows. Dry concentration measurements for CO, $CO_2$, and

$O_2$ were recorded continuously and combustion gases were collected periodically for gas chromatographic analysis. A stainless steel probe was used for gas extraction.

Gas concentrations, temperatures, and burning rates were averaged and plotted as functions of $Ah^{1/2}$ for a fuel loading of $\approx$ 3 kg. Five differently scaled enclosures were tested. Results for different sized enclosures were collapsed in terms of $Ah^{1/2}/S^{1.9}$. Note the good agreement with the findings of Gross and Robertson who used an exponent of 2 for S [37],[38]. The results seem to fall into four regimes based on changes in the observed concentrations, temperatures, and burning rates with increasing ventilation parameter. In regime I the temperatures were low (500-600 K). Measured concentrations of $O_2$ were roughly constant at 4%, while CO concentrations varied between 5% and 9%, and $CO_2$ attained values of $\approx$ 18%. As the ventilation parameter was increased the temperature began to rise, finally reaching an asymptotic value of $\approx$ 1000 K. Initially, both CO and $O_2$ concentration dropped as the temperature increased with CO falling to a minimum of 3% while $O_2$ became less than 1%. For the largest ventilation parameters investigated, the $O_2$ remained quite low while CO again rose to $\approx$ 11%. The data treatment chosen by Tewarson makes comparisons with other work difficult, but it is clear that there are cases for low upper-layer temperatures where fuel and $O_2$ exist together. There are also higher temperature cases where very high levels of CO are found and $O_2$ concentrations are extremely low.

Tewarson has attempted to use the wood fire data discussed above, along with other findings, to quantify the generation efficiencies of chemical compounds in enclosure fires burning wood cribs [40]. This work includes a plot of measured CO yields (g CO generated/g fuel consumed) plotted as a function of the air-to-fuel stoichiometric ratio (note this equivalence ratio is the inverse of that used elsewhere in this report). A curve is included suggesting that for rich conditions $\approx$ 0.2 grams of CO are produced per gram of wood consumed, and that the values fall off with increasing air-to-fuel ratio (decreasing equivalence ratio). Tewarson notes that many data points lie considerably above this curve and attributes these observations to the following possibilities: 1) experimental errors, 2) sampling within flames, 3) ventilation effects due to wood-crib construction, 4) gas temperatures and residence time of oxygenated compounds in the upper layer, and 5) changes in wood decomposition mechanisms.

Beyler has also considered this study and has argued that the high readings for CO concentrations noted by Tewarson can all be attributed to fires in which sampling directly in the fire was likely [18]. Beyler concludes that the dependence he finds for CO concentrations on the global equivalence ratio from his hood experiments [18] is in agreement with the enclosure fire results when measurements in fire zones are excluded. This point will be discussed further in Section V.C.3

Morikawa and Yanai [41] have studied fires in a reduced-full scale room (a cube 2 m on a side) in which controlled amounts of air were injected into the enclosure at the bottom and fire gases were exhausted at the top. The fuel loads consisted of a range of plastics, fabrics, and wood typical of a furnished room. Temperature measurements at various locations in the enclosure along with concentrations (wet) of combustion gases sampled in the exhaust duct by a stainless-steel probe are reported. Observed temperatures were generally in the 900-1300 K range.

70

Concentrations of CO initially rose quite rapidly and generally reached values greater than 7%, and occasionally values greater than 10% were observed. Oxygen concentrations tended to clump around values of 5% during the intense burning phase. These results are somewhat different than others discussed earlier in that high concentrations of fuel molecules and oxygen were observed simultaneously in fires where the combustion gases were at high temperatures.

It is difficult to reach conclusions from the studies summarized above. Not only were the ventilation arrangements vastly different in each case, the test methodologies and measurements techniques were highly variable. There were clearly cases for which the measurements were dependent on sampling procedures. Even though there are a number of tests reported, it must be concluded that the production of CO by these solid-fueled fires in enclosures is not well characterized.

B.    **Recent Experiments Performed at the Virginia Polytechnic Institute and State University (VPISU)**

An experimental system especially designed to test the ability of the GER to correlate the concentrations of gas species observed in the upper layer of a fire localized within an enclosure has been developed at VPISU [42],[43]. Figure 34 shows a schematic for their experimental system. The enclosure dimensions are width:length:height = 1.2 m : 1.5 m : 1.2 m. It was designed such that air, which was naturally entrained into the fire, entered only through an inlet duct and distribution plenum located beneath the floor of the enclosure. Slits along the length of the floor allowed the air to enter the enclosure. This

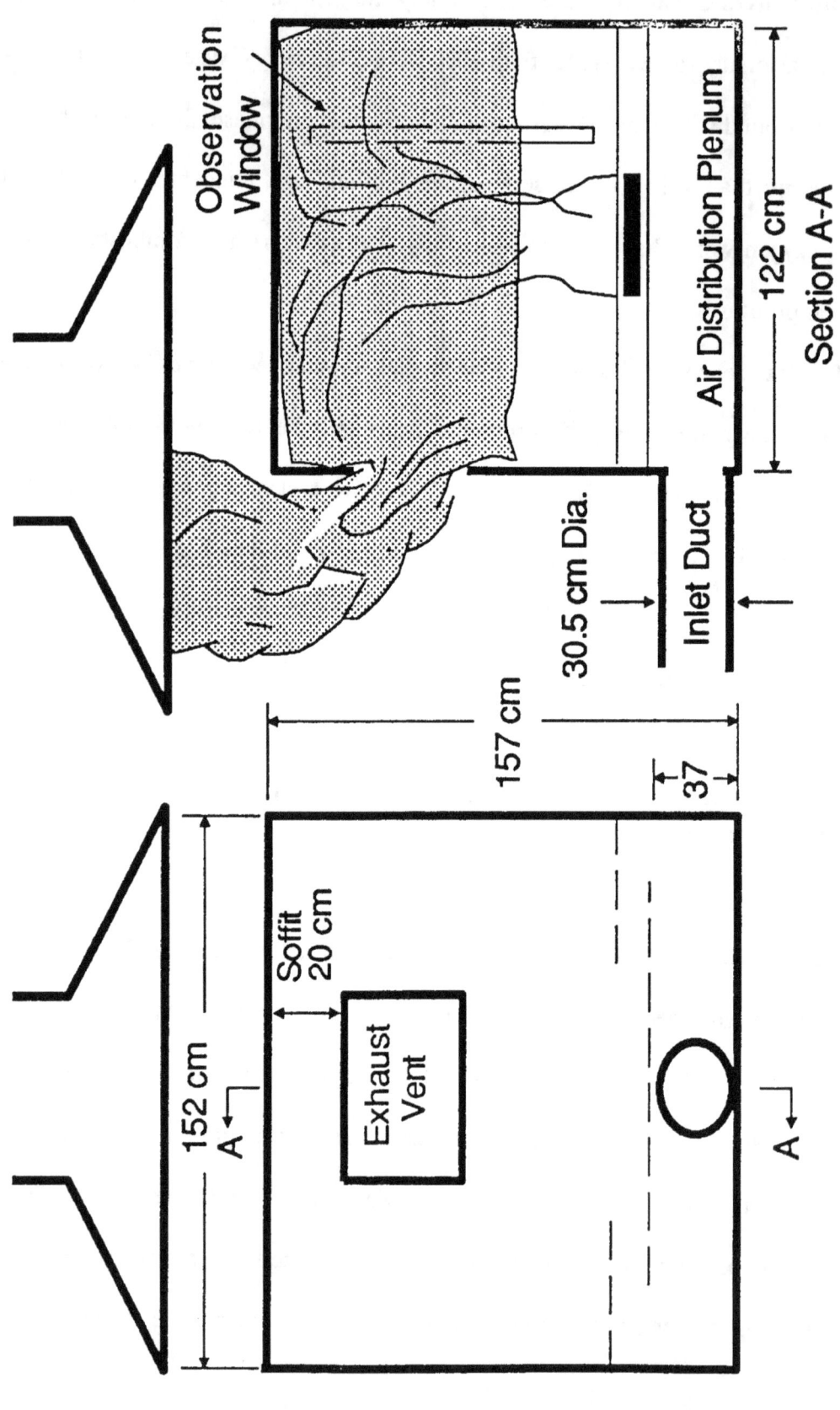

Figure 34. Schematic of the experimental system developed by Roby and coworkers at VPISU for the investigation of combustion gas generation in enclosure fires having two well-defined layers. Figure provided by Dan Gottuk.

configuration had two important advantages. First, it was possible to measure the air flow in the inlet duct and therefore directly characterize the air flow rate into the fire. Second, the distribution plenum ensured that air entered the lower layer slowly over a wide area. As a result, the two-layer structure of the gases within the enclosure was strongly stabilized.

The exhaust for the fire was a single window located in one of the long sides of the enclosure. Careful tests showed that there was no air entrained into the enclosure through this window, and only combustion products exited the window. The window size was varied to provide different ventilation conditions. Products exhausted from the enclosure were collected in a large exhaust hood and withdrawn through a duct.

By measuring the mass loss rate of the fuel and the rate of air mass inflow, it was possible to determine $\phi_p$ simply by dividing their ratio by that required for stoichiometric burning. During their experiments it was found that most fires eventually achieved a pseudo-steady state and that $\phi_p \approx \phi_g$.

Gases in the upper layer were sampled with an uncooled stainless steel probe. By moving the probe tip to various locations it was demonstrated that species concentrations were independent of position for points outside of the fire plume. Concentration measurements were recorded for $CO$, $CO_2$ and $O_2$. Temperatures were recorded at several locations using thermocouples.

Four fuels were studied: hexane, PMMA, wood, and a polyurethane foam containing 45% by weight of inert filler. Hexane was burned in pans, PMMA as a sheet which melted, wood as fuel-surface-controlled cribs, and the foam as a sheet. All fuels were burned on a load cell and the mass loss as a function of time was recorded.

73

Figure 35 shows examples of the gas and equivalence ratio measurements as a function of time for a hexane burn. Note that concentrations are reported on a wet basis, despite the fact that water was removed during the measurement. In order to correct for water removal, the assumption was made that the molar ratio of $H_2O$ and $CO_2$ at any $\phi_p$ was the same as for stoichiometric burning. A distinct quasi-steady burning behavior can be seen.

By averaging over the quasi-steady burning periods, concentrations as a function of $\phi_p$ were obtained. Since the fires are in a pseudo-steady state, $\phi_p \approx \phi_g$. Figures 36 and 37 show results for $O_2$ and CO for the four fuels. It can be seen that the $O_2$ decreases with increasing $\phi_p$ for lean conditions and becomes very nearly zero under rich conditions for all four fuels. Note that natural ventilation limited the tests for wood and polyurethane to $\phi_p < 2$. At the same time, CO concentrations appear to remain very low until $\phi_p \approx 1$, at which point they begin to rapidly rise to an asymptotic value which is different for each fuel with the order being hexane < wood $\approx$ polyurethane < PMMA. Observed concentrations of CO for rich conditions range from 2.6 to 4.2% The CO concentrations for the PMMA fires have more scatter than observed for the other fuels. This is unexplained at the present time.

Beyler investigated three of the fuels studied at VPISU ($C_6H_{14}$, wood, and PMMA) in his hood experiments [16],[17],[18]. Figures 38-40 compare the findings for CO generation in the enclosure and hood experiments for these three fuels. Several points are important to note. The CO concentrations in both experiments have an "S" shape with low concentrations for small $\phi_p$. However, the two curves are offset in $\phi_p$ by $\approx 0.5$ with the hood results starting to increase at $\phi_p = 0.5$ and the enclosure results at $\phi_p = 1.0$. For the

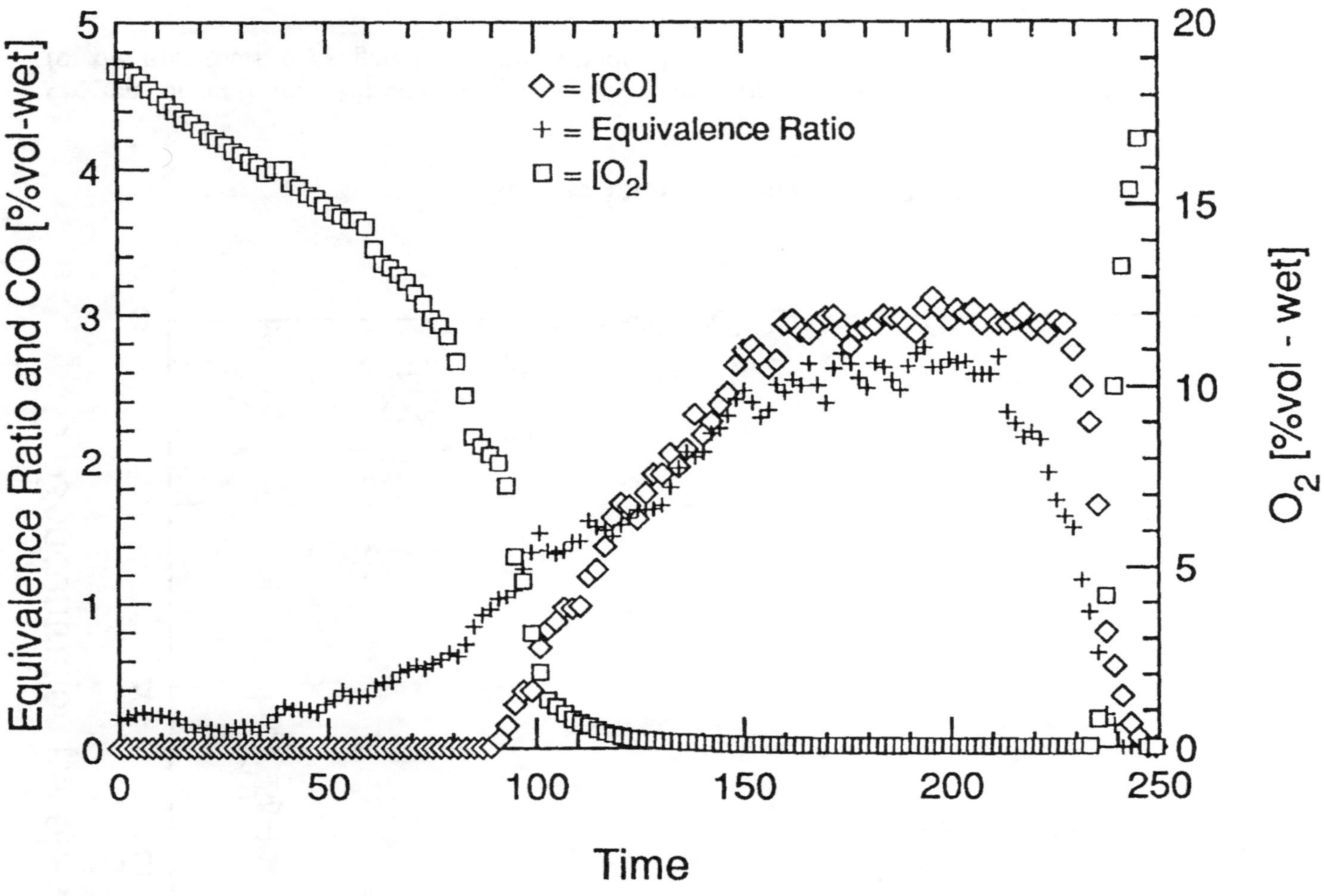

<u>Figure 35.</u>    Equivalence ratio and CO and $O_2$ volume percents (wet) are shown as a function of time for a hexane-fueled fire in the VPISU facility. The development of the fire and the attainment of a pseudo-steady state are evident. Figure provided by Dan Gottuk.

75

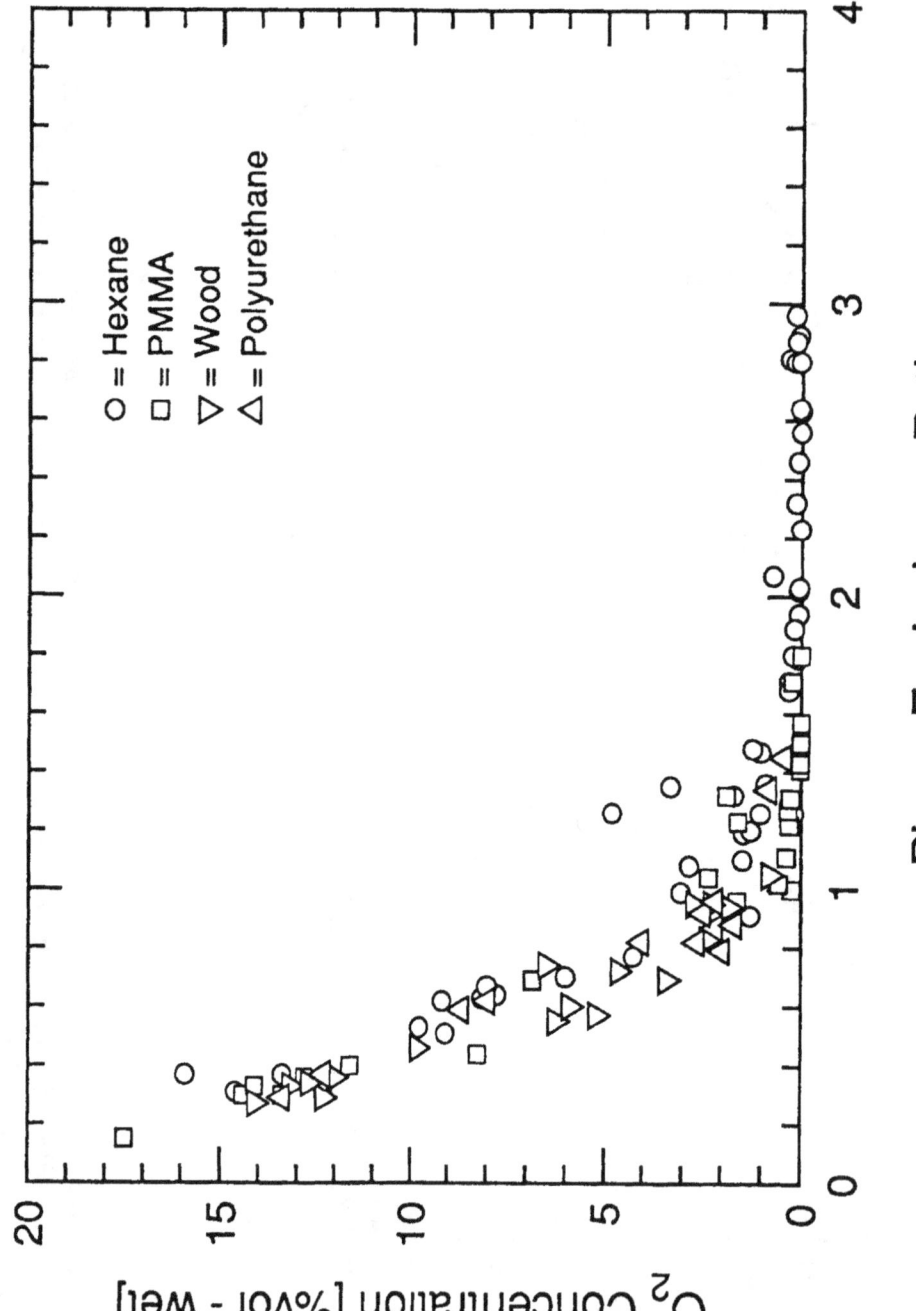

Figure 36.    Oxygen concentrations observed during steady-state burning in the VPISU facility are plotted as a function of $\phi_p$ for the four fuels investigated.  Figure provided by Dan Gottuk.

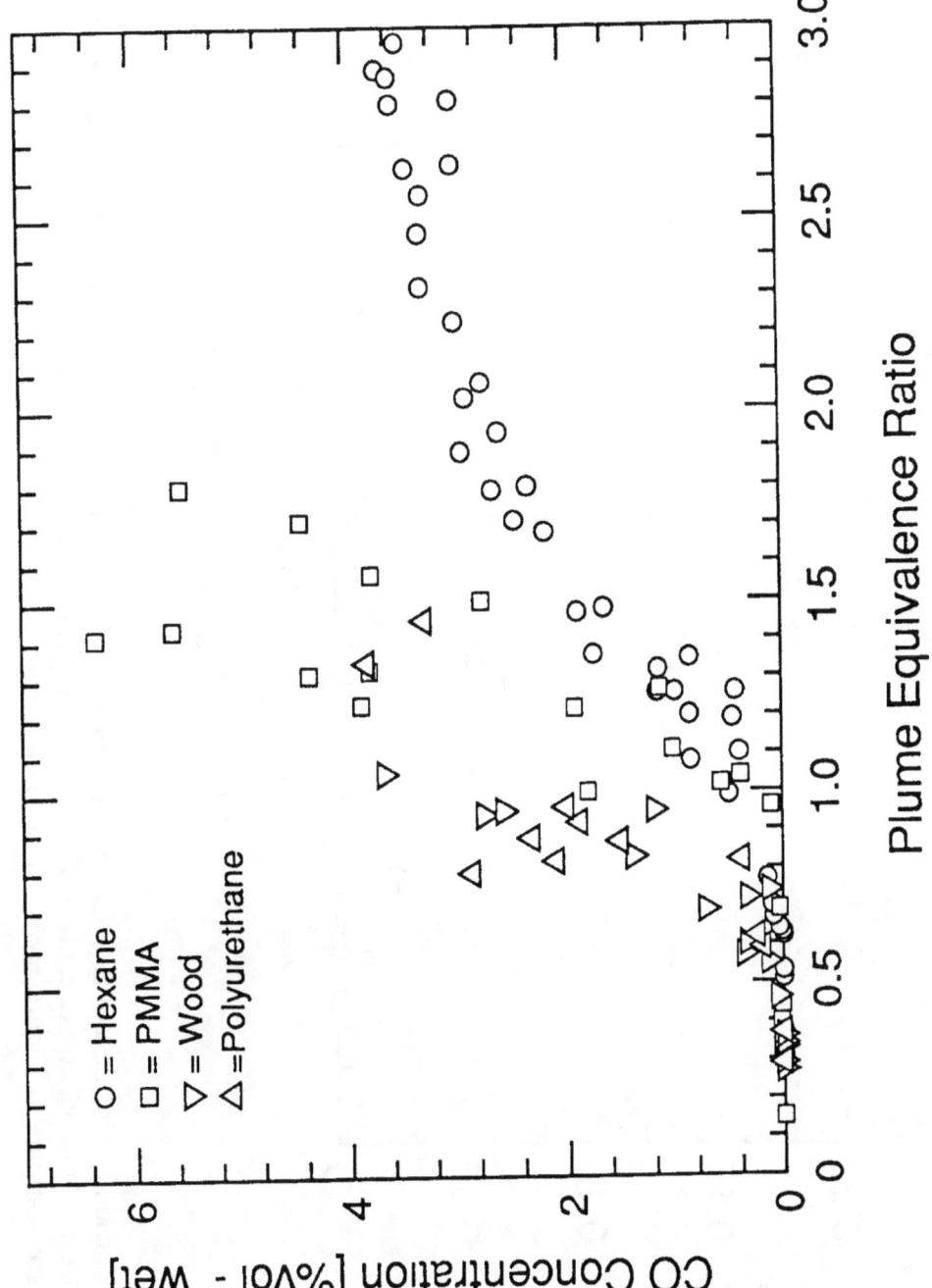

Figure 37.   Carbon monoxide concentrations observed during steady-state burning in the VPISU facility are plotted as a function of $\phi_p$ for the four fuels investigated.   Figure provided by Dan Gottuk.

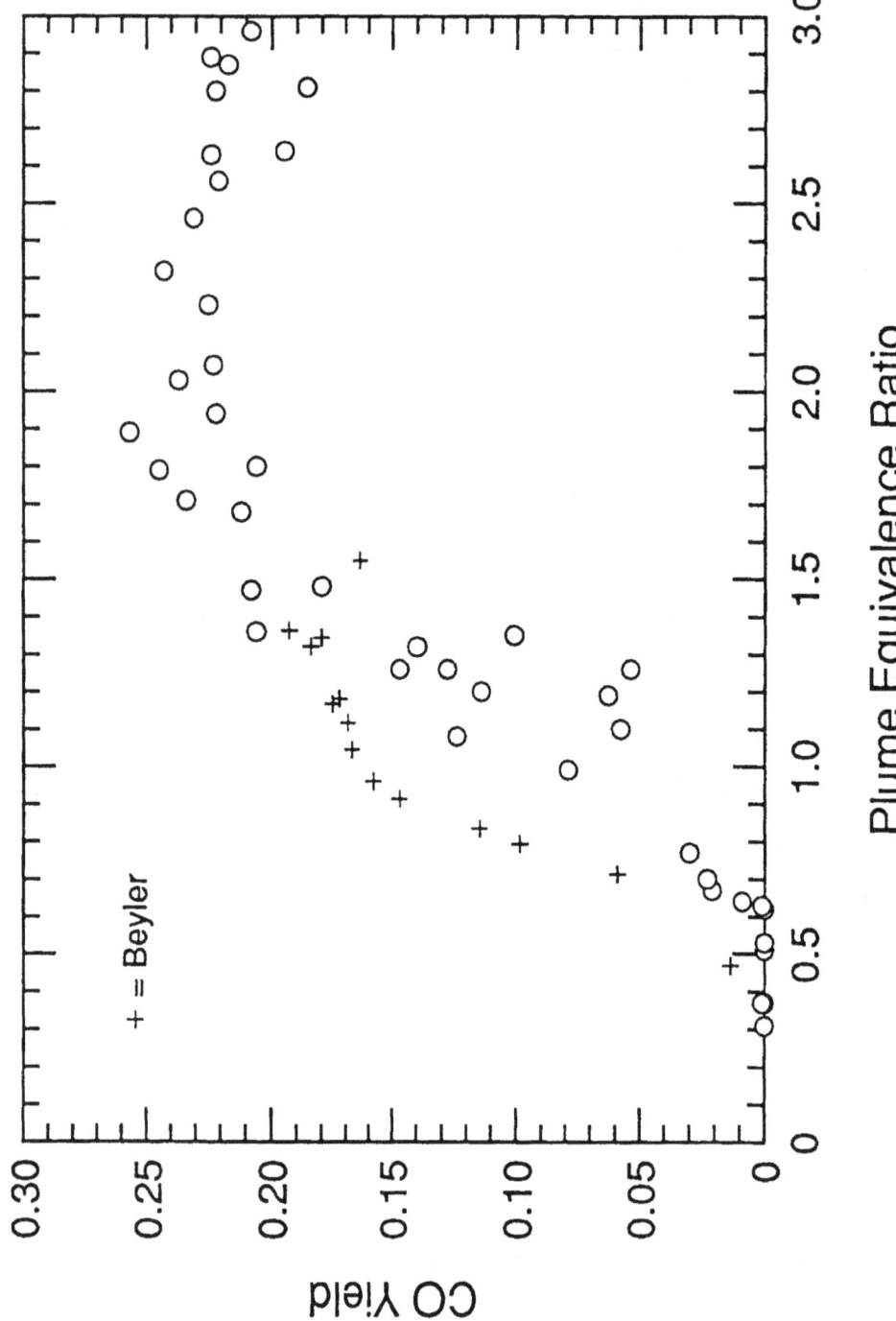

<u>Figure 38.</u>   Carbon monoxide yields (g CO produced/g fuel consumed) as a function of $\phi_p$ for hexane fuel are plotted. Data from the hood experiments of Beyler [16], [17] and the enclosure experiments of Roby and coworkers [42],[43] are compared. Figure provided by Dan Gottuk.

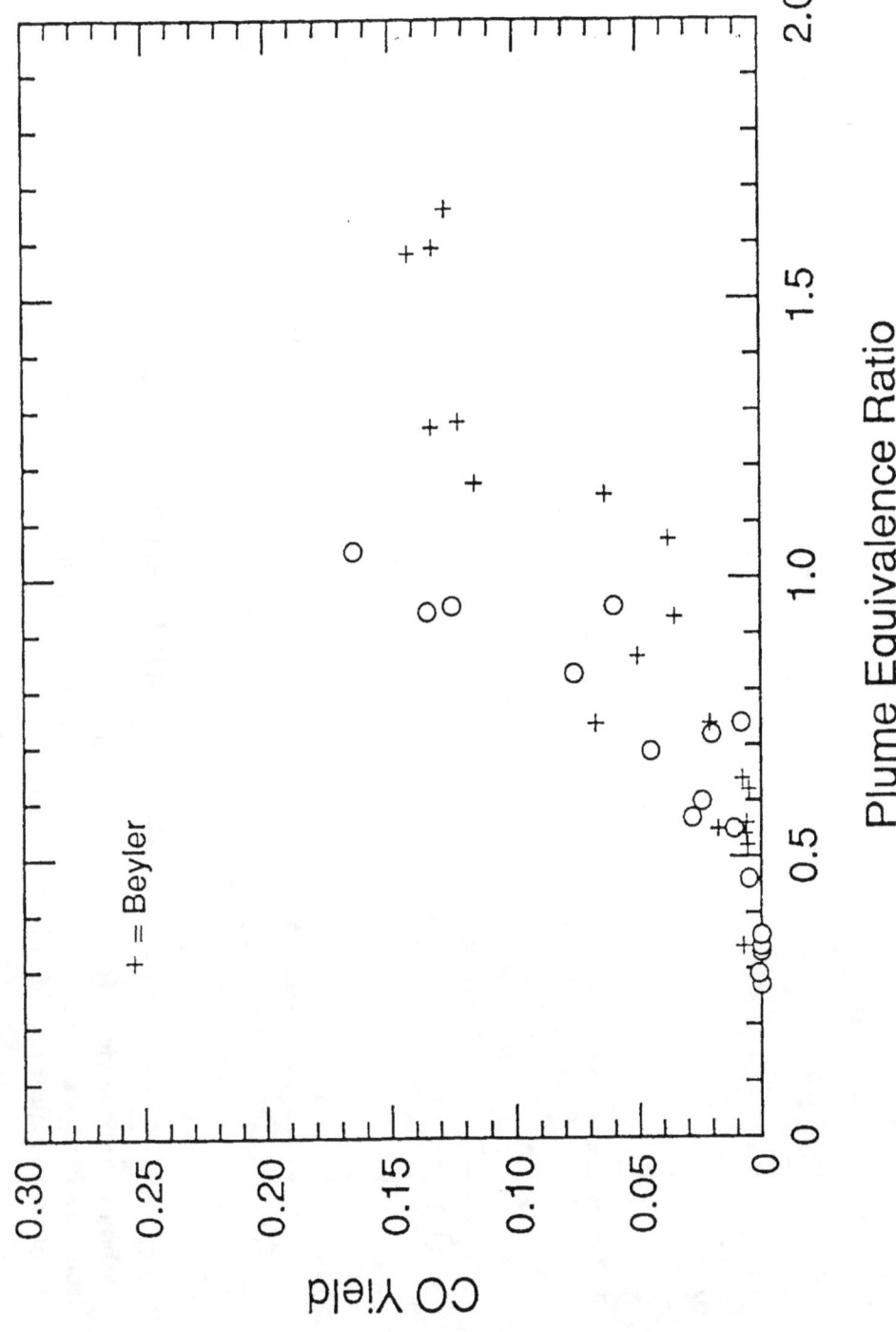

Figure 39.    Carbon monoxide yields (g CO produced/g fuel consumed) as a function of $\phi_p$ for wood fuel are plotted. Data from the hood experiments of Beyler [18] and the enclosure experiments of Roby and coworkers [42],[43] are compared. Figure provided by Dan Gottuk.

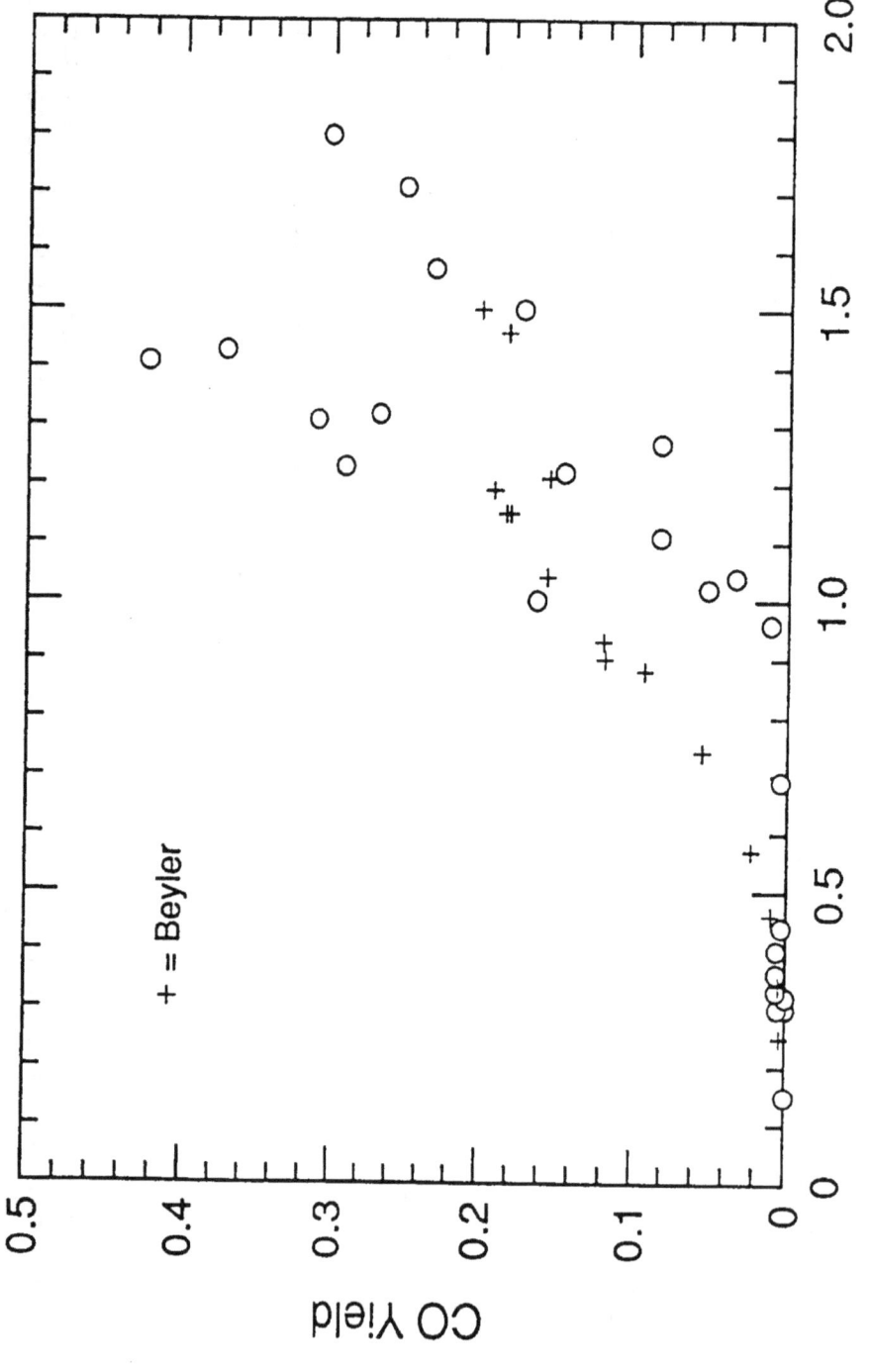

Plume Equivalence Ratio

Figure 40.  Carbon monoxide yields (g CO produced/g fuel consumed) as a function of $\phi_p$ for PMMA fuel are plotted.  Data from the hood experiments of Beyler [18] and the enclosure experiments of Roby and coworkers [42],[43] are compared.  Figure provided by Dan Gottuk.

rich fires, temperatures in the upper layers within the enclosure were all greater than 800 K (in a range of 800-1170 K). These are considerably higher than in the hood experiments (470-800 K). The shift in the S curve with increasing temperature is consistent with the observations and conclusions of Morehart and coworkers [23],[25] as discussed above.

Morehart's data was inconclusive concerning the effects of the shifts in reaction behavior at higher temperatures. The findings shown in figures 38-40 indicate that the CO yields attain asymptotic values for high $\phi_p$ in both the hood and enclosure experiments. The asymptotic yields appear to be slightly higher for the enclosure experiments. Comparisons between the hood and enclosure data for the yields of $O_2$ and $CO_2$ are inconclusive, but suggest that increasing the upper-layer temperature has the net result of depleting the oxygen, which results primarily in the generation of $CO_2$. Comparisons of Morehart's [23],[25] and Toner's [22] hood data lead to the same conclusion--higher hood temperatures resulted in the generation of both CO and $CO_2$, but the generation of $CO_2$ was favored. A similar comparison of Morehart's and Beyler's [16],[17] results suggests that the additional reaction generated primarily $CO_2$.

The results discussed in this section support the idea that the composition of gases measured in the low-temperature hood experiments is changed when a upper layer reaches higher temperatures. Though the data are limited, it can also be argued convincingly that the yields of CO are only slightly increased at higher temperatures in the hood and enclosure experiments for high $\phi_g$, and that the high temperature dependence of CO formation on $\phi_g$ can be estimated by simply shifting the S curve obtained from the low-temperature hood experiments to higher $\phi_g$ by $\approx$ 0.5. Note, that as discussed earlier, this requires that the relative species generation rates of the fire plume be altered as the temperature is raised.

As shown by Morehart's results [23],[25] and discussed above, a range of temperatures and upper-layer residence times are to be expected over which the simplification of two S curves will fail. The findings of the VPISU study suggest that this temperature range is centered somewhere near 800 K and is narrow. For temperatures typical of fully developed enclosure fires, the high temperature curves for a given fuel should be adequate.

## C. Summary of Recent Experimental and Modeling Investigations of Reduced-Scale Enclosure (RSE) Fires at NIST

A systematic investigation of the behavior of enclosure fires focused on the formation of CO is being performed at NIST [44]. The findings up to the present time are summarized here.

### 1. Experimental Findings for Natural-Gas Fires in a Reduced-Scale Standard Room

The enclosure used in these studies is a 2/5 scale model of a room which has been proposed for use in both ASTM [45] and ISO [46] standard fire tests. The model is referred to as the reduced-scale enclosure (RSE). It has dimensions of 0.98 m (width) x 1.46 m (length) x 0.98 m (height). The standard full-scale room contains a single doorway in the short wall which is 0.76 m wide and 2.03 m high. The standard scaling law

[47],[48],[49] for enclosure ventilation normalized by burning rate is expressed as $Ah^{1/2} = CG^2$ where A is the area, h the height of the vent, G is the size ratio for geometrically similar enclosures, and C is a constant. This scaling law is the same as used by Gross and Robertson [37],[38]. The resulting doorway dimensions for the RSE are 0.48 m x 0.81 m.

The RSE was located under a furniture calorimeter [50] which collected the combustion gases exiting the doorway and allowed measurements of heat release rate (HRR) as well as the relative concentrations of CO and $CO_2$ for positions far removed from the enclosure. Standard instrumentation for RSE fire characterization consisted of arrays of thermocouples and two banks of analyzers containing a paramagnetic analyzer for oxygen and non-dispersive infrared (NDIR) analyzers for CO and $CO_2$. A vertical array of aspirated thermocouples (thermocouple tree) located at the doorway along with a differential pressure measurement across the doorway allowed an estimate of the entrainment of air through the doorway using two approaches described in the literature [51],[52]. The approach developed by Janssens and Tran [52], which uses a mass balance approach to locate the neutral plane at the doorway, yielded more accurate results [44] and was used for the measurements of global $\phi$ which are discussed later in the text. A second thermocouple tree was located in the rear of the enclosure.

For a few tests, additional instrumentation was utilized. A total hydrocarbon analyzer was used to measure unburned fuel. An instrument, dubbed the $\phi$-meter, which was designed and built at NIST [53], was used to measure the local equivalence ratio of the combustion gases extracted from the enclosure. The $\phi$-meter is a prototype instrument which uses a catalyst to convert any excess fuel and products of incomplete combustion to $CO_2$ and $H_2O$. For these conditions, a single measurement of the $O_2$ concentration allows the local equivalence ratio to be calculated [53].

Gas measurements were made for dried and filtered samples extracted from various locations in the upper and lower layers within the enclosure. For most experiments, two 0.94 cm inside-diameter stainless-steel tubes were used to sample gases. Generally, one of these tubes would be passed into the upper layer through the enclosure wall at positions near the front or rear of the enclosure. The second probe was usually a long probe which was inserted through the doorway and could be positioned at the any desired location. Measurements using this second probe were recorded at various locations in the upper and lower layers of the fire. For some fires, multiple locations were sampled during a single burn. Figure 41 shows the locations within the RSE for which probe measurements were made.

Due to concerns about the possibility of reactions in the probes (see earlier discussion) a water-cooled stainless-steel probe was constructed and used in some tests. Reaction was a particular concern in the case of the long probe when used for measurements in the upper layer. This probe usually passed through a flame zone located at the layer interface, as well as extensive distances in the hot upper-layer gases. Comparisons of measurements in comparable, but different, fires using the long uncooled and cooled probes for the same sampling position were in good agreement, suggesting that no significant reactions were occurring in the uncooled probes.

The fuel for the initial fire tests was natural gas supplied by a 15 cm diameter burner centered in the RSE and located 15 cm above the floor. This fuel was chosen since it is

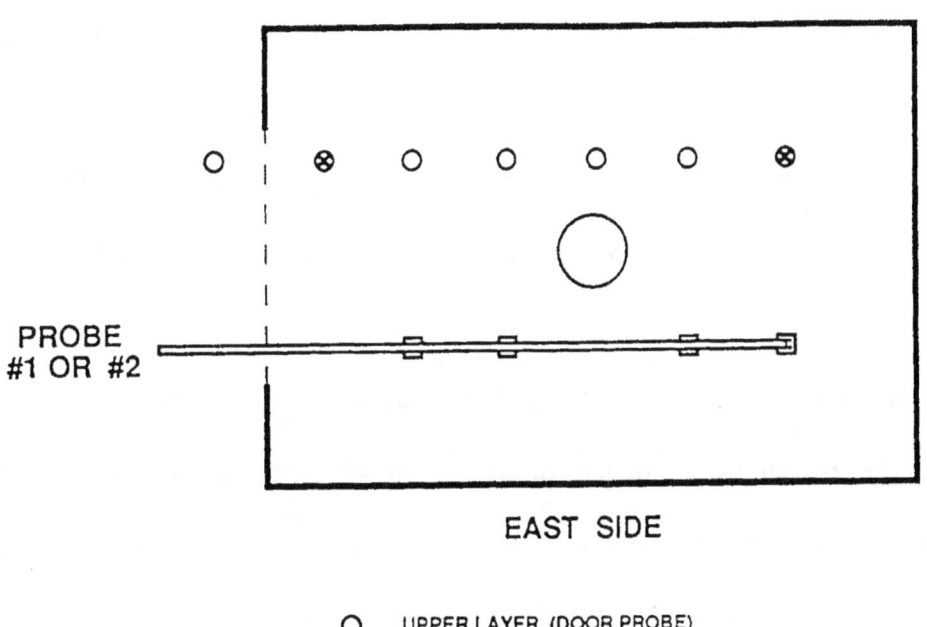

EAST SIDE

○ UPPER LAYER (DOOR PROBE)

✕ UPPER LAYER (WALL PROBE)

☐ LOWER LAYER (DOOR PROBE)

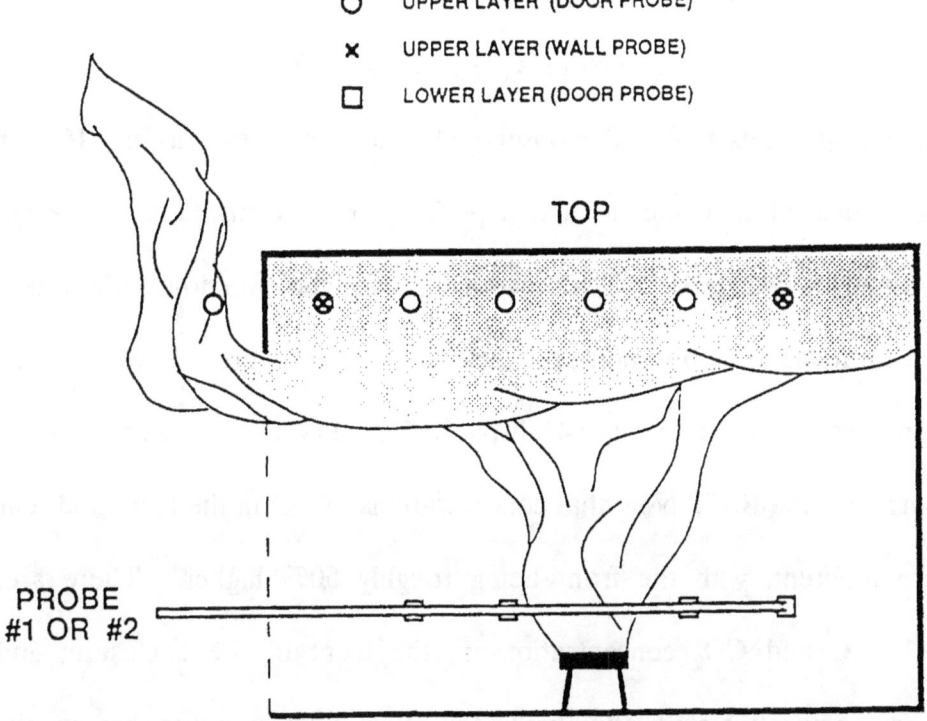

Figure 41. A schematic of the reduced-scale enclosure showing locations where probes were placed to extract gases. Generally, two probes were employed with one being a wall probe and the second a long probe passing through the doorway. For some fires the doorway probe was moved to several positions.

85

easily metered and controlled and because the studies of Morehart [23],[24],[25] and Toner [22] provided an extensive data base of hood-type experimental results for comparison purposes. Nominal heat release rates for the fires (assuming complete combustion) ranged from 10 to 670 kW. Gas flow rates were maintained constant for the 15-20 min duration of a fire.

The time behaviors of the concentrations of $O_2$, CO, and $CO_2$ in the front and rear of the enclosure are shown in figure 42 for a 500 kW fire. This fire is significantly underventilated. The concentrations have been corrected for water removal (i.e., they are wet concentrations and calculated water concentrations are included in the figure) using the same approximation as Roby and coworkers [42],[43]; i.e., it is assumed that water and $CO_2$ are in the same proportions as would be observed for stoichiometric burning. It can be seen from the figure that, after a brief induction period, the concentrations of the gases are relatively constant or change slowly. By averaging over the pseudo-steady-state burning periods, concentrations of combustion products typical of each fire size were obtained.

The time records shown in figure 42 display behaviors which are typical of underventilated fires burned in the RSE. Note that concentrations of CO in the front and rear of the enclosure are different, with the front being roughly 50% higher. There are larger fluctuations in CO and $CO_2$ concentrations in the front of the enclosure, and these fluctuations have a negative correlation with each other suggesting that they are the result of changes in the reaction behavior. For both locations, the $O_2$ concentrations decreased rapidly and remained much less than 0.5% during the test.

86

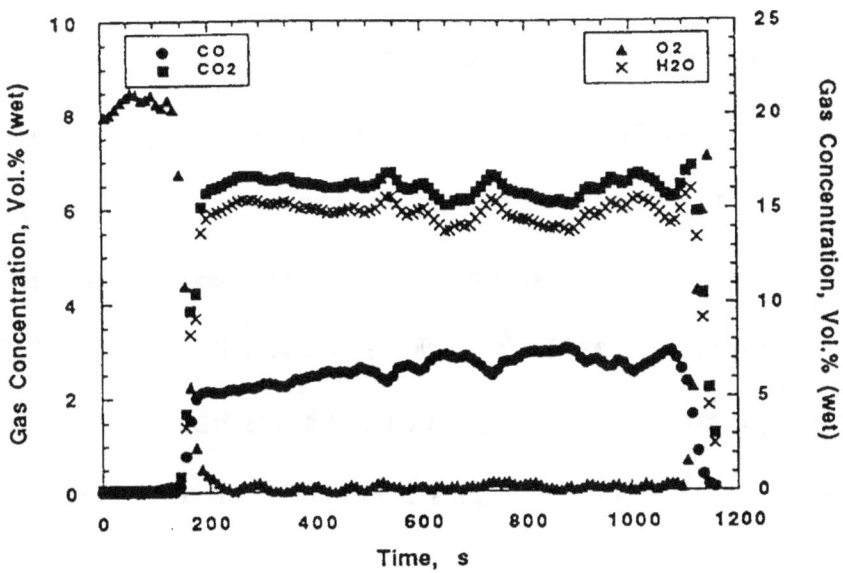

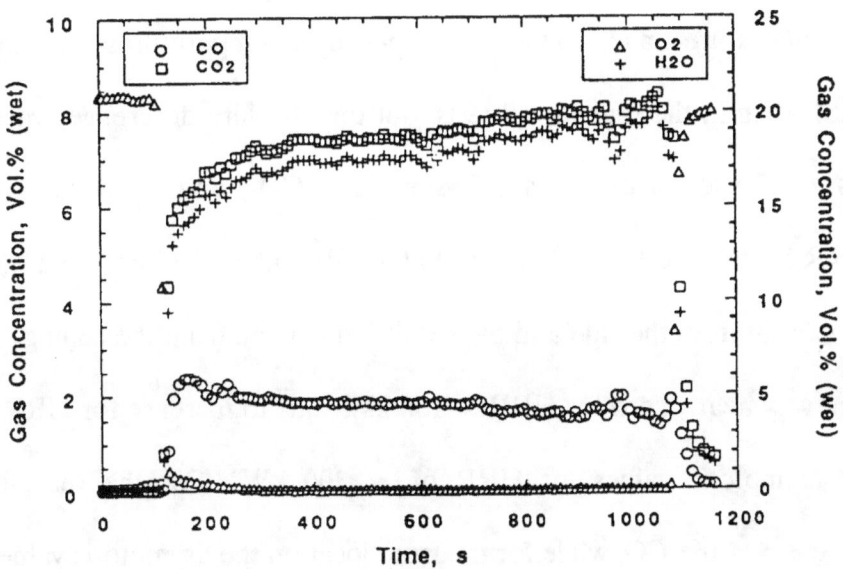

Figure 42. Carbon monoxide, carbon dioxide, oxygen, and calculated water concentrations (wet percents) are plotted as a function of time for a 500 kW natural gas fire in the RSE. Measurements in the upper layer in the front (solid symbols; 10 cm from front wall and ceiling, 29 cm from side wall) and rear (open symbols; 29 cm from rear and side walls, 10 cm from ceiling) are shown.

Measurements of temperature as a function of time for various heights near the front of the enclosure are shown in figure 43 for a 500 kW fire. The upper layer is located $\approx$ 60 cm from the floor and has a relatively uniform temperature varying between 1100 K and 1300 K. The lower layer is heated by radiation, but is considerably cooler and quite non-uniform. Figure 44 shows pairs of time records of thermocouple measurements for upper-layer locations in the front and rear of the enclosure for two 400 kW fires. The two front thermocouples were located at the same position within the RSE while the rear thermo-couples were located at two different positions as indicated. For each general position the measurements are in good agreement, but the front of the enclosure is $\approx$ 1300 K while the rear is about 300 K cooler. The good agreement between the two fires indicates the reproducibility of the fires is good, and that for the height of the rear measurements, temperature gradients are small. Generally, upper-layer temperatures were higher in the front of the RSE than in the rear for all tests, but the absolute differences varied.

Figures 45-47 show averaged steady-state CO, $CO_2$, and $O_2$ concentrations as a function of HRR for two locations--29 cm from the side wall and 10 cm from the front wall and ceiling and 29 cm from the side and back walls and 10 cm from the ceiling. Concentrations of CO are very nearly zero for HRR < 100 kW, start to increase for HRR > 100 kW, and approach asymptotic values for HRR of $\approx$ 300 kW. For the rear location the asymptotic value is $\approx$ 2.0% CO, while for the front location the asymptotic value is $\approx$ 3.0% or roughly 50% larger. Carbon dioxide concentrations initially rise with HRR and are identical for front and rear sampling at HRR < 200 kW. For HRR rates > 200 kW the $CO_2$ concentrations level off and begin to decrease slightly. Values in the rear of the

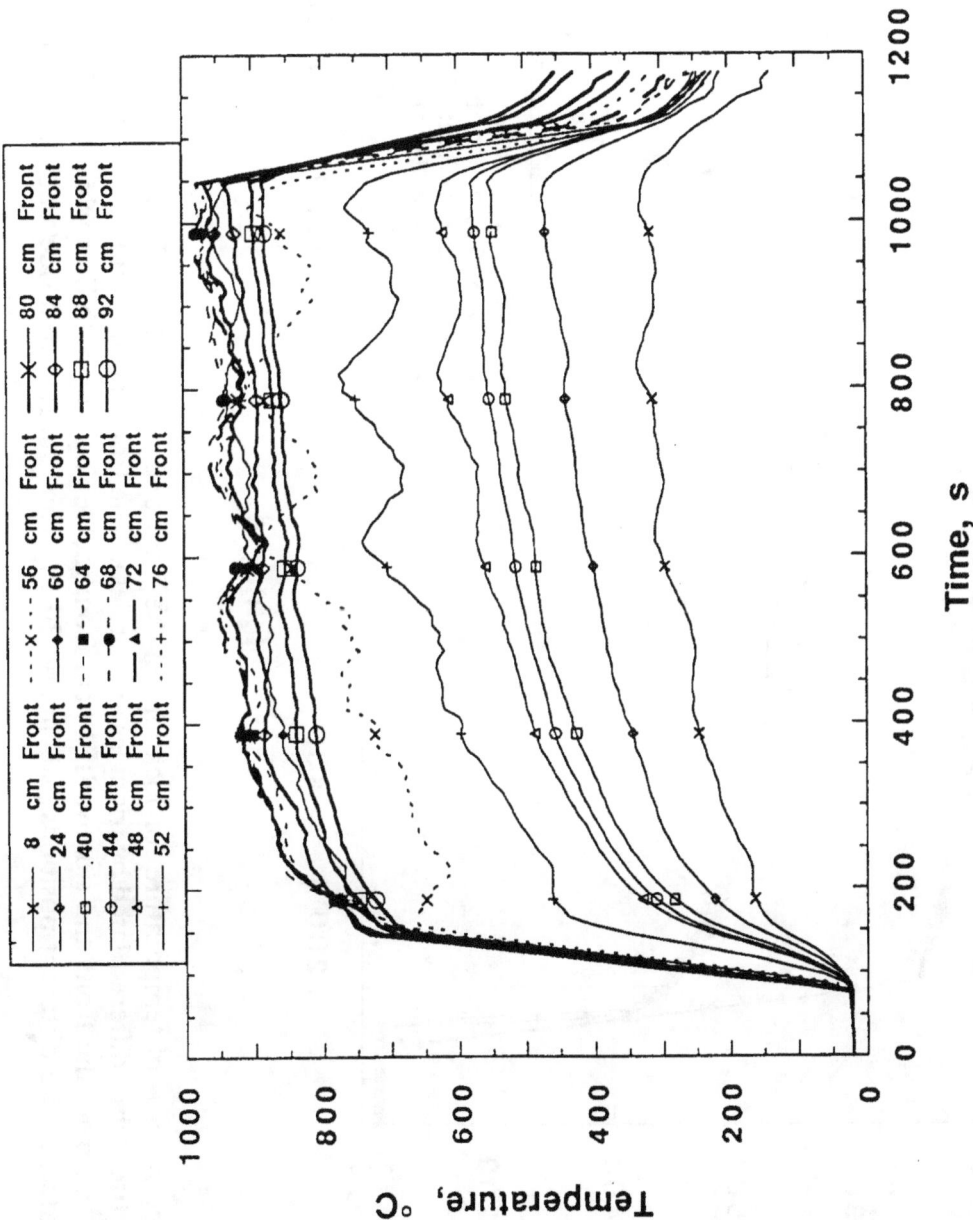

Figure 43.    Temperature measurements using a vertical array of thermocouples are shown as a function of time for a 500 kW natural gas fire. The thermocouple tree was placed 20 cm from the front and side walls of the RSE.

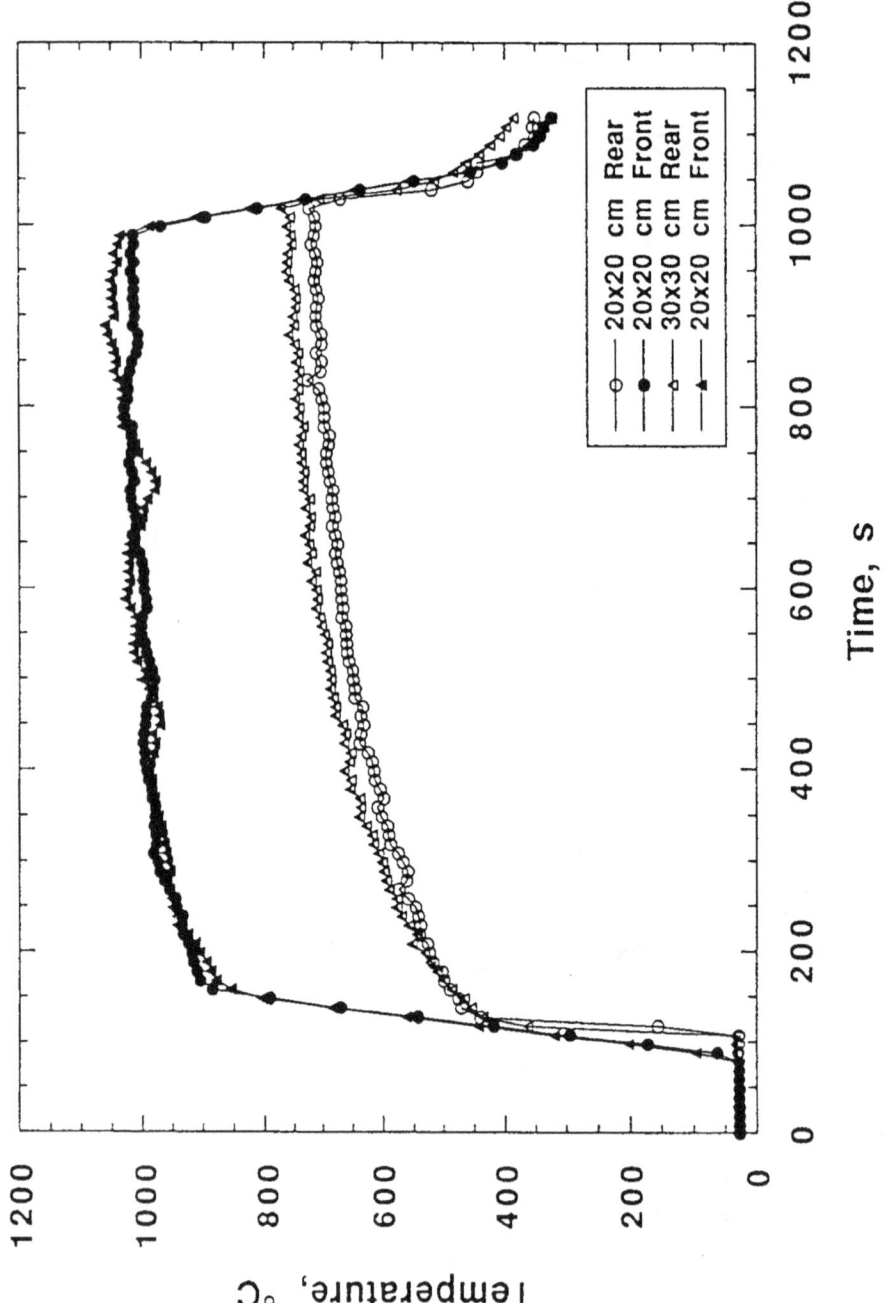

Figure 44. Time records of temperature for locations in the front and rear of the RSE are shown for two 400 kW natural gas fires (fires differentiated by triangles and circles). The front thermocouples were located at the same position (20 cm from the front and side walls, 18 cm from the ceiling) for both fires while the rear thermocouples were located at two different positions (20 cm and 30 cm from the rear and side walls, 18 cm from the ceiling).

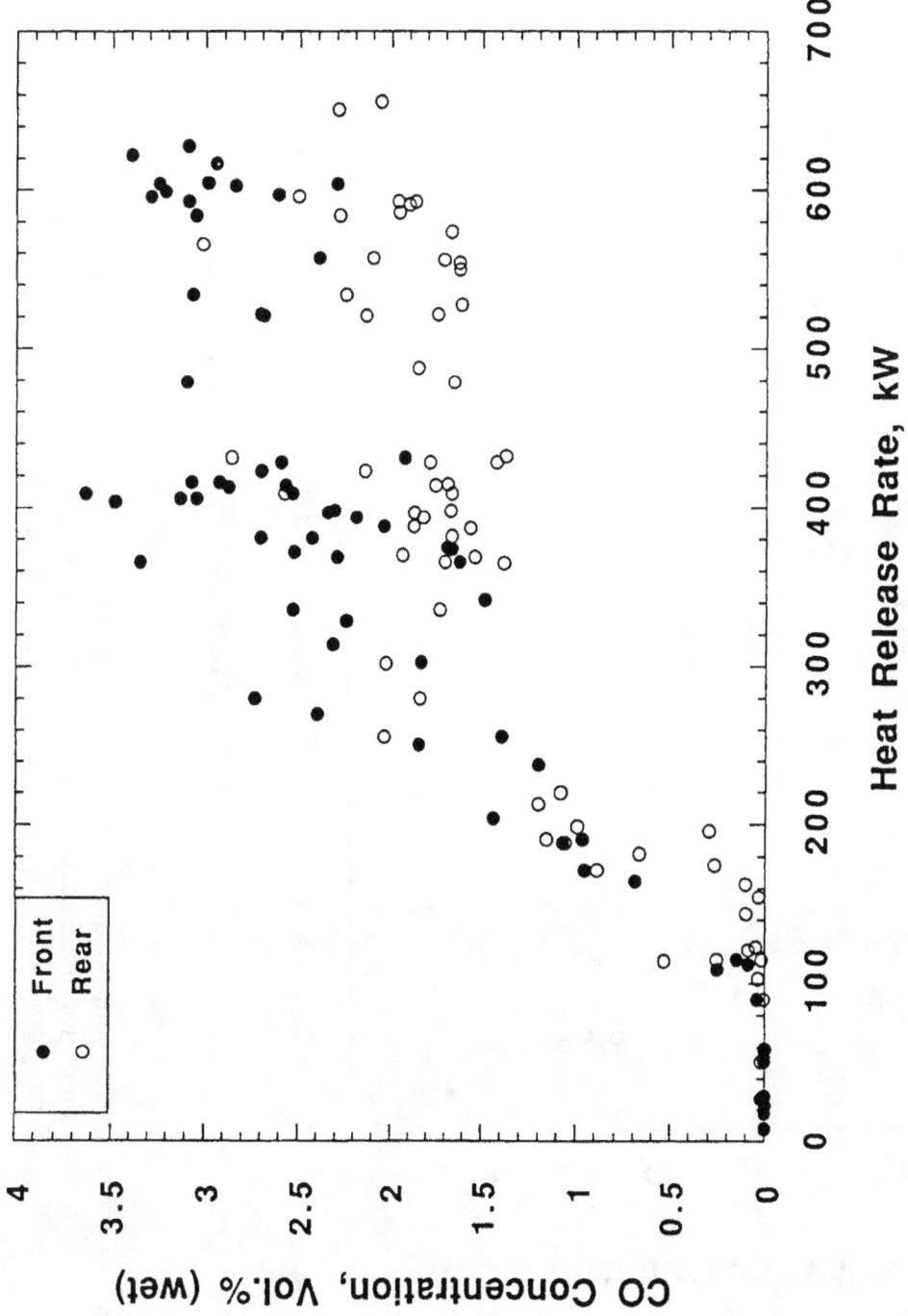

<u>Figure 45.</u>    Carbon monoxide concentrations observed during pseudo-steady state burning of natural gas fires in the RSE plotted as a function of nominal heat release rate. Measurement positions were in the front (10 cm from front wall and ceiling, 29 cm from side wall) and rear (29 cm from side and rear walls, 10 cm from ceiling) of the RSE.

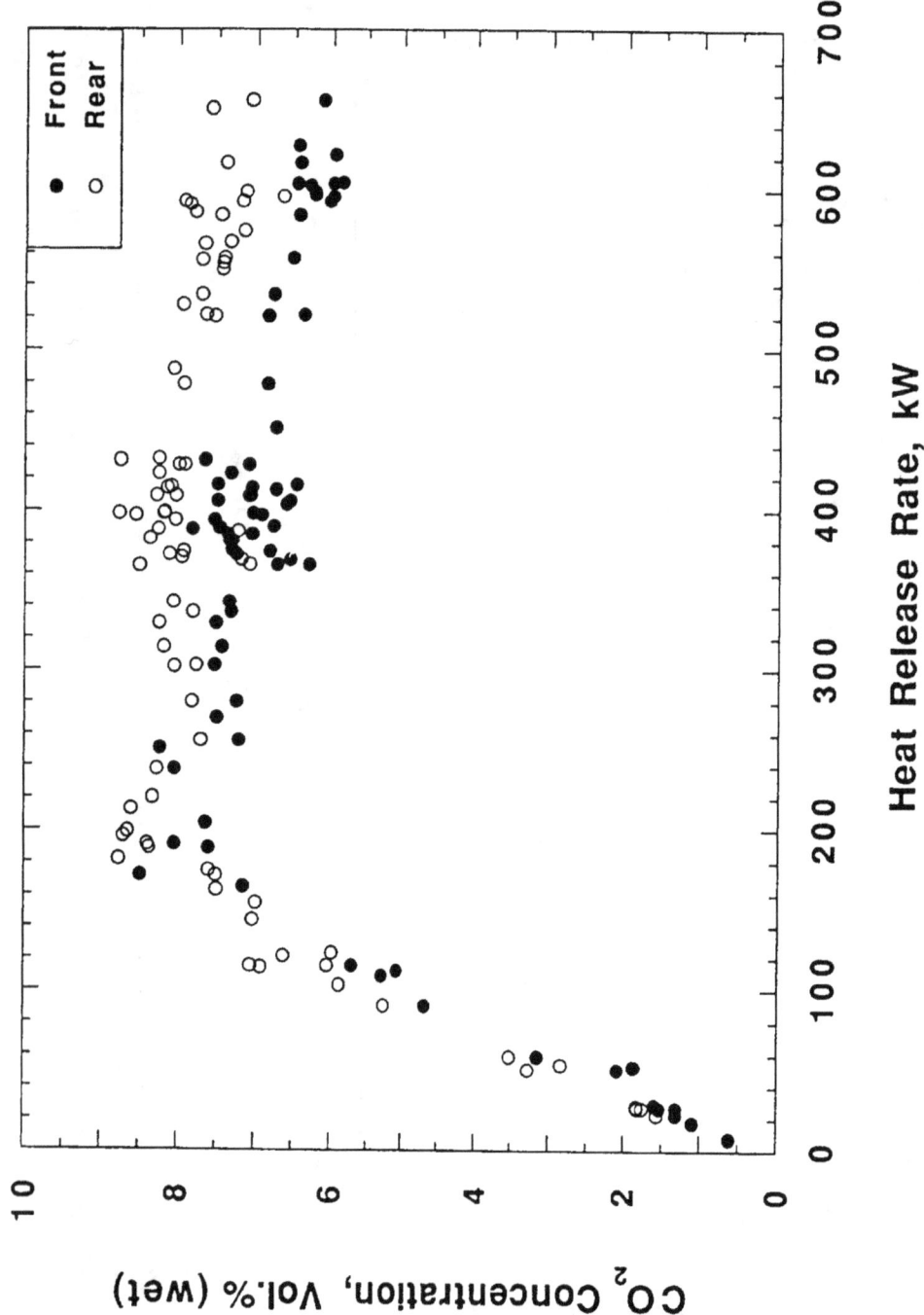

Figure 46. Carbon dioxide concentrations observed during pseudo-steady state burning of natural gas fires in the RSE plotted as a function of nominal heat release rate. Measurement positions were in the front (10 cm from front wall and ceiling, 29 cm from side wall) and rear (29 cm from side and rear walls, 10 cm from ceiling) of the RSE.

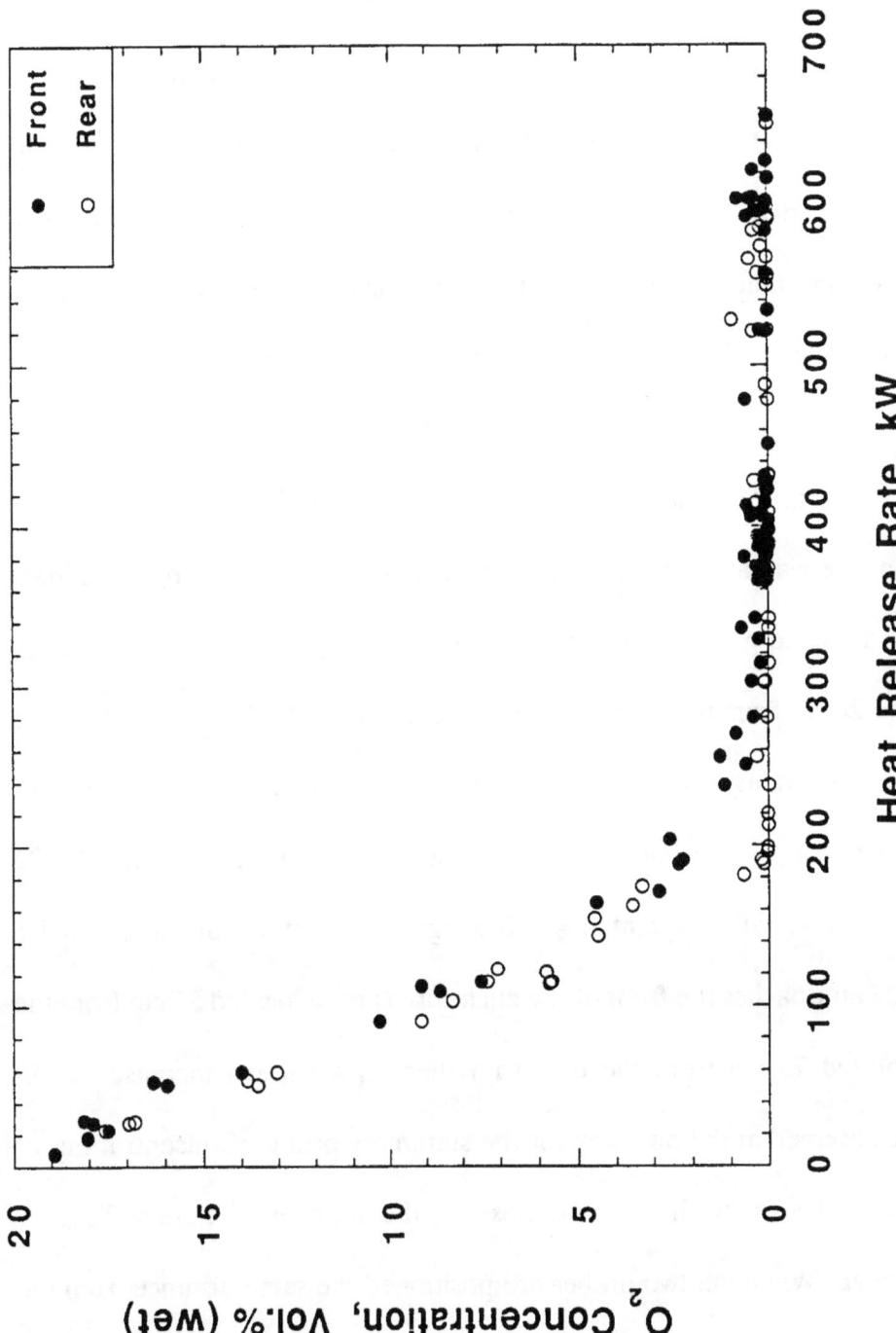

Figure 47.  Oxygen concentrations observed during pseudo-steady state burning of natural gas fires in the RSE plotted as a function of nominal heat release rate.  Measurement positions were in the front (10 cm from front wall and ceiling, 29 cm from side wall) and rear (29 cm from side and rear walls, 10 cm from ceiling) of the RSE.

enclosure are slightly higher than in the front. Oxygen concentrations fall from near-ambient values for small HRR fires to values less than 1% for HRR > 300 kW. Both front and rear locations have very similar behaviors except over the narrow range from 200 to 300 kW, where oxygen concentrations in the front are slightly above zero while those in the rear are very close to zero. The lack of $O_2$ in the upper layer at high HRRs, and the fact that the relative amounts of CO and $CO_2$ observed in the front and rear show inverse behavior clearly indicate that combustion is more efficient in the rear of the enclosure where the concentration ratio of $CO/CO_2$ is lower. The more efficient combustion in the rear is observed despite the significantly lower temperatures at this position.

Figure 48 shows measurements of CO concentration in a 600 kW fire for a probe located 29 cm from a side wall and 10 cm from the front wall and ceiling, along with data recorded by scanning a second probe parallel to the side wall along a line positioned 29 cm from the side wall and 20 cm from the ceiling. The concentrations of CO measured for the stationary probe grow to an asymptotic value of $\approx$ 3.5%, consistent with the earlier measurements. For locations in the rear of the enclosure, the second probe records CO concentrations which are relatively constant at $\approx$ 1.8%, again consistent with earlier results. However, as the probe approaches the front of the enclosure (probe located 51 cm from the doorway, burner centered 73 cm from the doorway) there is a sudden increase in CO concentration to that observed at the doorway for the stationary probe. Concentrations of CO observed by the movable probe then fall off closer to the doorway, but are still higher than observed in the rear. When the two probes are positioned the same distance from the doorway, the mobile probe, which is lower in the layer, records a lower CO concentration.

94

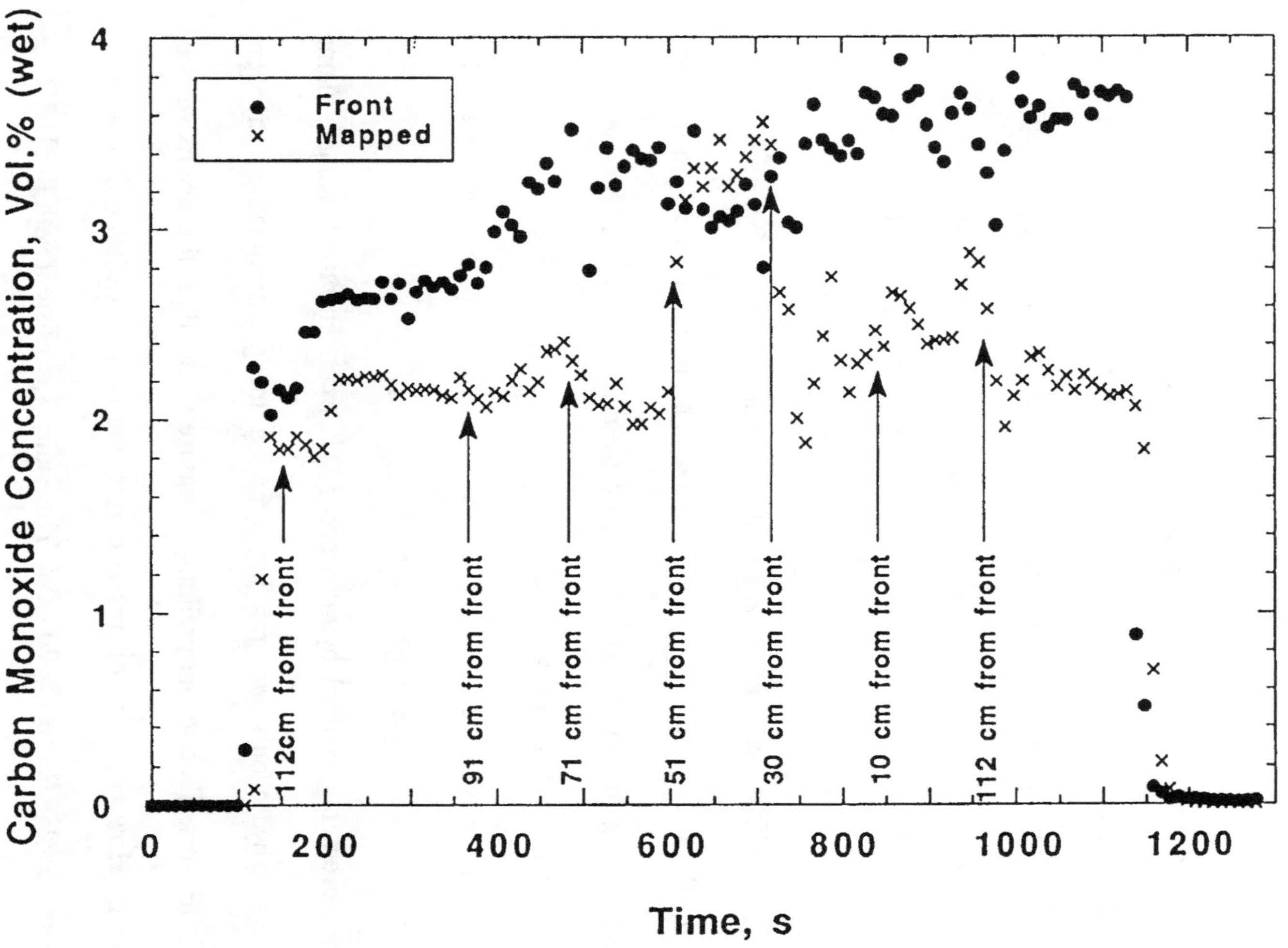

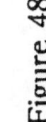

Figure 48. Time records of carbon monoxide concentration are shown for two probes sampling a 600 kW natural gas fire burning in the RSE. One probe is stationary and is positioned in the front of the enclosure 29 cm from the side wall and 10 cm from the front wall and ceiling. The second probe is moved during the fire along a horizontal line located 29 cm from the side wall and 20 cm from the ceiling.

95

These observations are attributed to variations in CO concentration with vertical position in the upper layer as one moves toward the front of the enclosure. Note that the final position of the movable probe for the measurements in figure 48 is in the rear of the enclosure and reproduces the initial measurement quite well.

In the previous paragraph it was suggested that CO concentration varies with vertical position in the upper layer. Figure 49 shows measurements for a stationary probe and a scanned probe located 29 cm from the side wall and 10 cm from the front wall. The stationary probe is 10 cm from the ceiling, and the height of the scanned probe is varied as indicated. It can be seen that the scanned and stationary probes record nearly identical concentrations for distances from the ceiling which are less than 10 cm. As the distance is increased, however, the CO concentration falls well below that observed by the stationary probe. Interestingly, the CO concentration seems to reach a minimum $\approx$ 25 cm from the ceiling before increasing again. At the largest distance from the ceiling the CO concentration drops rapidly as the probe is passing through the layer interface. These observations show that the upper layer is not well mixed. The variations of CO with height suggest that reactions are taking place within the upper layer.

Limited measurements of total hydrocarbon (THC) concentration using a flame ionization detector and hydrogen using grab-bag sampling and gas chromatography were recorded for the RSE fires. Due to the preliminary nature of the data the results will not be discussed in detail. However, it is worthwhile noting that values of the THC concentration began to increase for HRR rates > 200 kW. This value is consistent with the HRR for which the oxygen concentration approaches zero in the upper layer and for which

96

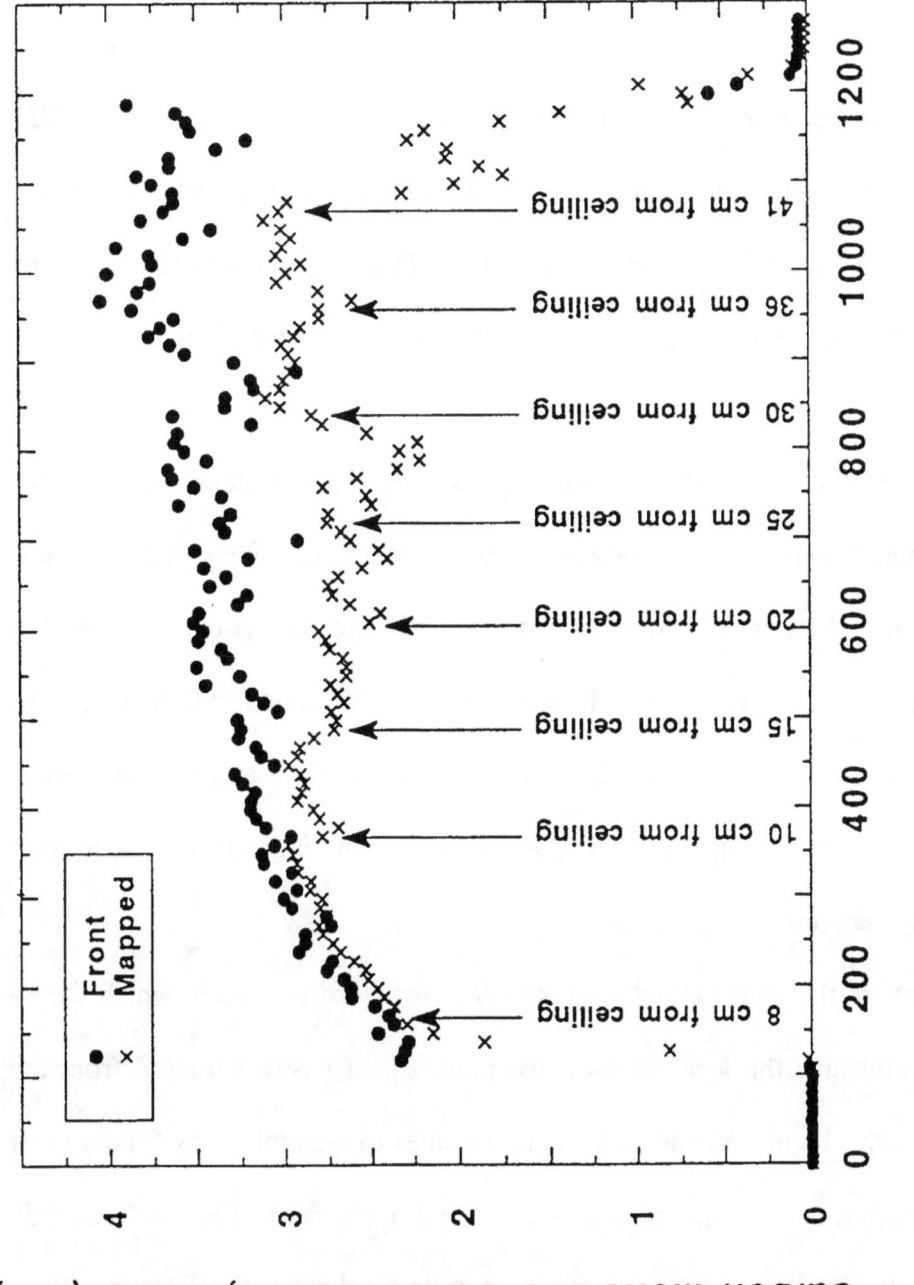

Figure 49. Time records of carbon monoxide concentration are shown for two probes sampling a 600 kW natural gas fire burning in the RSE. One probe is stationary and is positioned in the front of the enclosure 29 cm from the side wall and 10 cm from the front wall and ceiling. The second probe is moved during the fire along a vertical line located 29 cm from the side wall and 10 cm from the front wall.

concentrations of CO begin to increase. Hydrogen was detected in significant concentrations (1.1 - 2.9 dry volume percent) for fires having HRR in the range 250-600 kW. Though limited, the measurements show that hydrogen can be a significant product of incomplete combustion in enclosure fires. Hydrogen was also detected in the hood experiments.

Thus far, measured concentrations have been discussed only in terms of the HRR. Clearly, it is preferable to use a variable such as the GER. Unfortunately, the concentration measurements which were recorded are not sufficient to allow an equivalence ratio to be calculated. However, accurate entrainment rates into the enclosure were obtained using the analysis procedure developed by Janssens and Tran [52]. Using the derived air entrainment rates, the measured fuel flow rates, and by assuming the fires were in steady state, global equivalence ratios, denoted as $\phi_g$, could be calculated for the fires. Figures 50-52 show measured CO, $CO_2$, and $O_2$ concentrations plotted as functions of $\phi_g$. It can be seen that the $O_2$ levels fall to near zero for $\phi_g \approx 1.0$, and that the CO concentration begins to increase for roughly the same $\phi_g$. Since the CO and $O_2$ concentrations show the behaviors expected for a high temperature upper layer, these observations indicate that measured values of $\phi_g$ are fairly accurate.

To further confirm the above conclusion, several fires were burned in which the $\phi$-meter was used to determine the local equivalence ratio, $\phi_\ell$, of gases extracted from the front of the upper layer. Figure 53 shows measured values of $\phi_g$ and $\phi_\ell$ as functions of HRR. It can be seen that the two sets of data are in good agreement. Figures 54 and 55 show measured CO and $O_2$ concentrations in the front of the enclosure plotted as functions of $\phi_\ell$. It can be seen that the $O_2$ concentration also falls to zero for $\phi_\ell = 1$, and that the

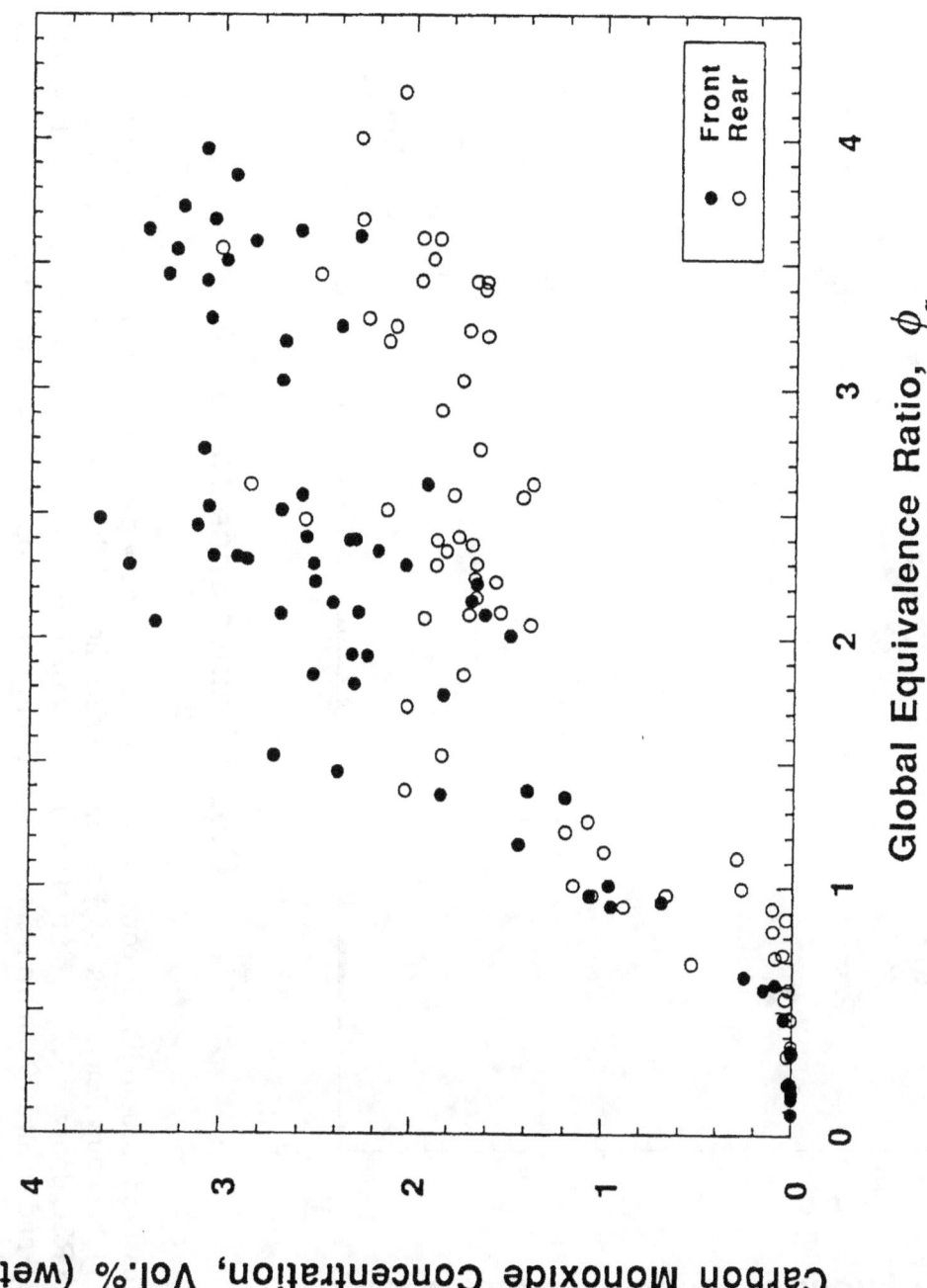

Figure 50. Measured carbon monoxide concentrations observed during pseudo-steady state burning are plotted as a function of global equivalence ratio ($\phi_g$) calculated using the fuel flow rate and an estimated rate of entrainment through the RSE doorway [52]. The fuel was natural gas. Results are shown for positions in the front (10 cm from front wall and ceiling, 29 cm from side wall) and rear (29 cm from side and rear walls, 10 cm from ceiling) of the RSE.

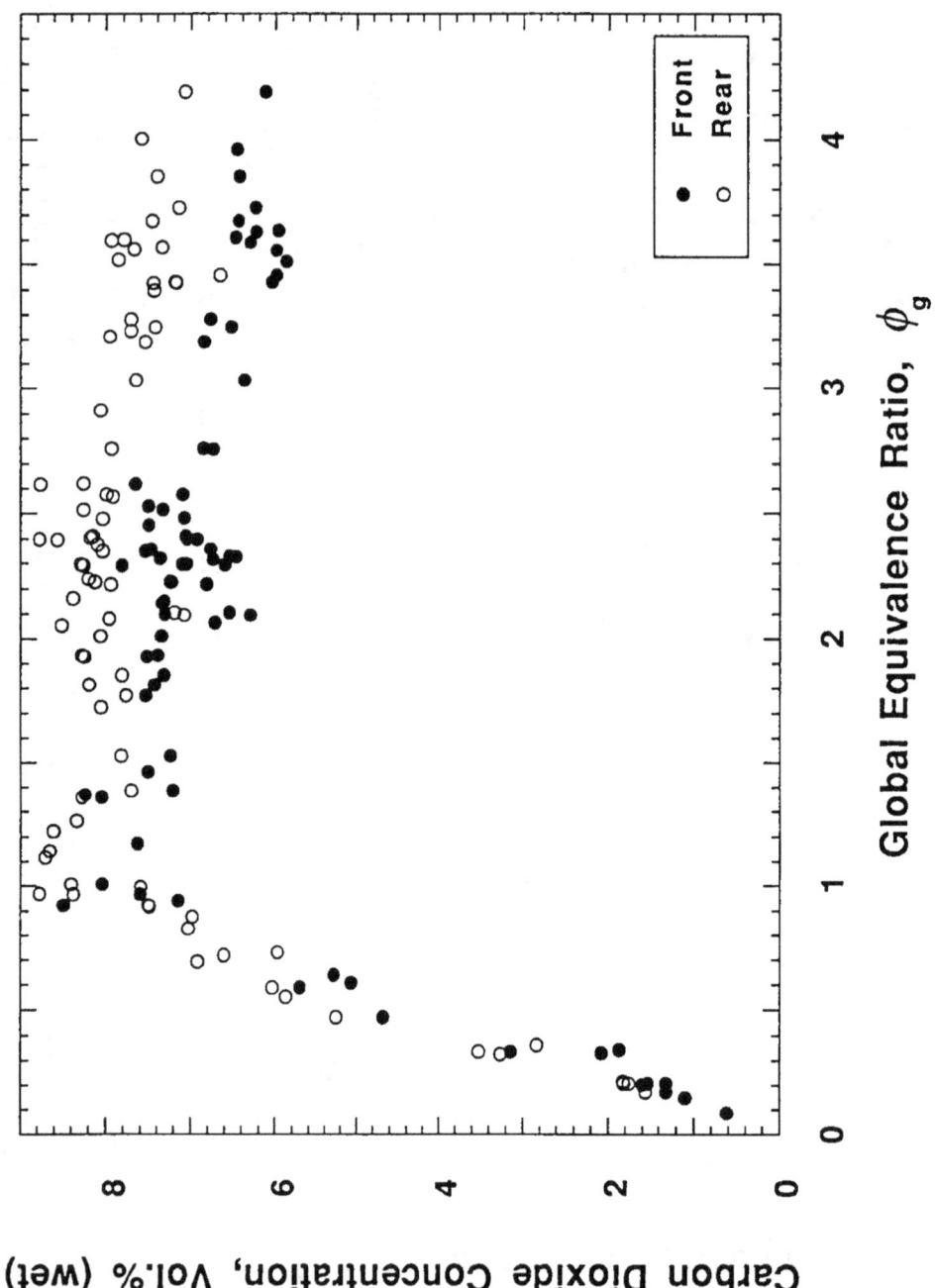

Figure 51.   Measured carbon dioxide concentrations observed during pseudo-steady state burning are plotted as a function of global equivalence ratio ($\phi_g$) calculated using the fuel flow rate and an estimated rate of entrainment through the RSE doorway [52]. The fuel was natural gas. Results are shown for positions in the front (10 cm from front wall and ceiling, 29 cm from side wall) and rear (29 cm from side and rear walls, 10 cm from ceiling) of the RSE.

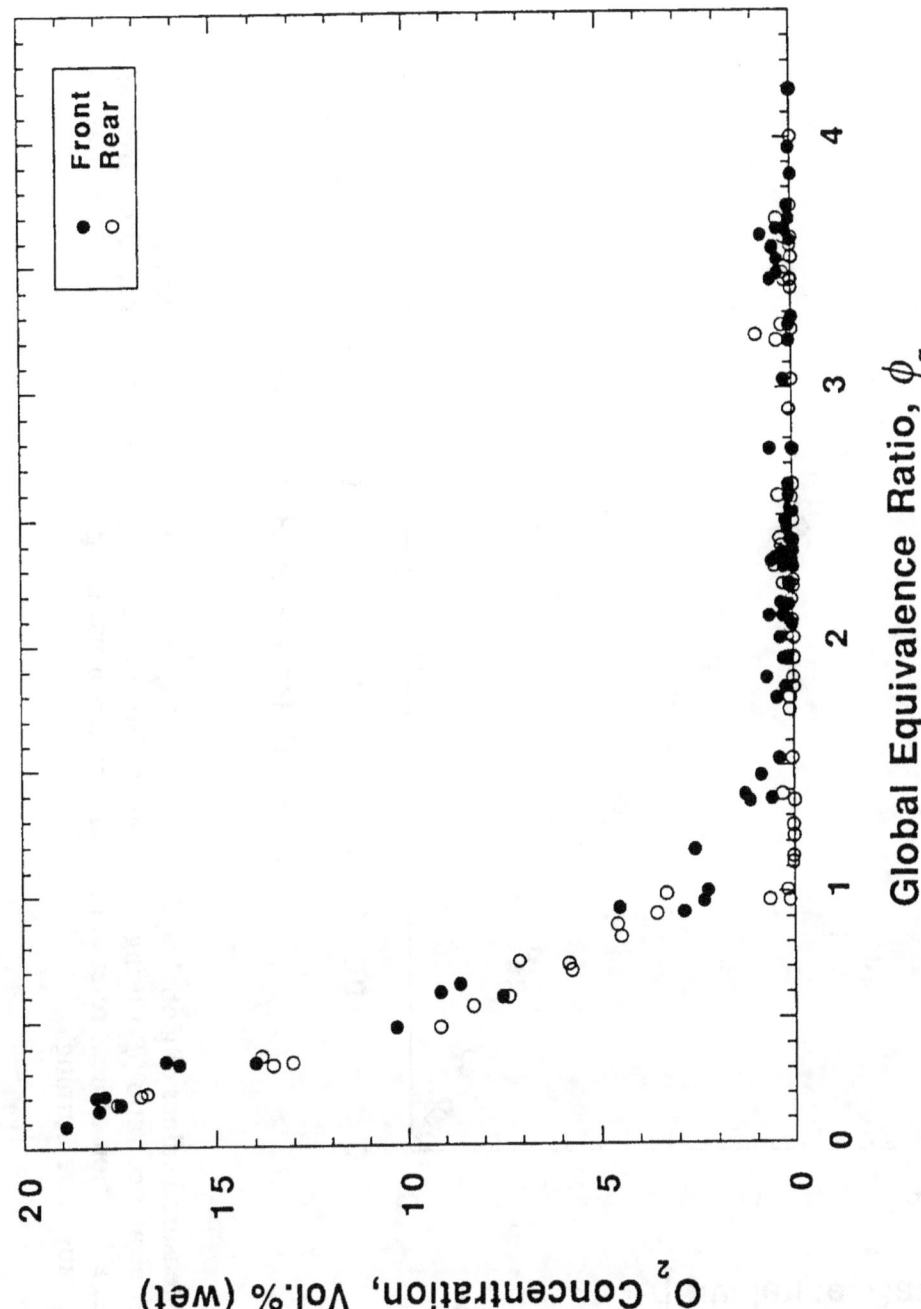

Figure 52. Measured oxygen concentrations observed during pseudo-steady state burning are plotted as a function of global equivalence ratio ($\phi_g$) calculated using the fuel flow rate and an estimated rate of entrainment through the RSE doorway [52]. The fuel was natural gas. Results are shown for positions in the front (10 cm from front wall and ceiling, 29 cm from side wall) and rear (29 cm from side and rear walls, 10 cm from ceiling) of the RSE.

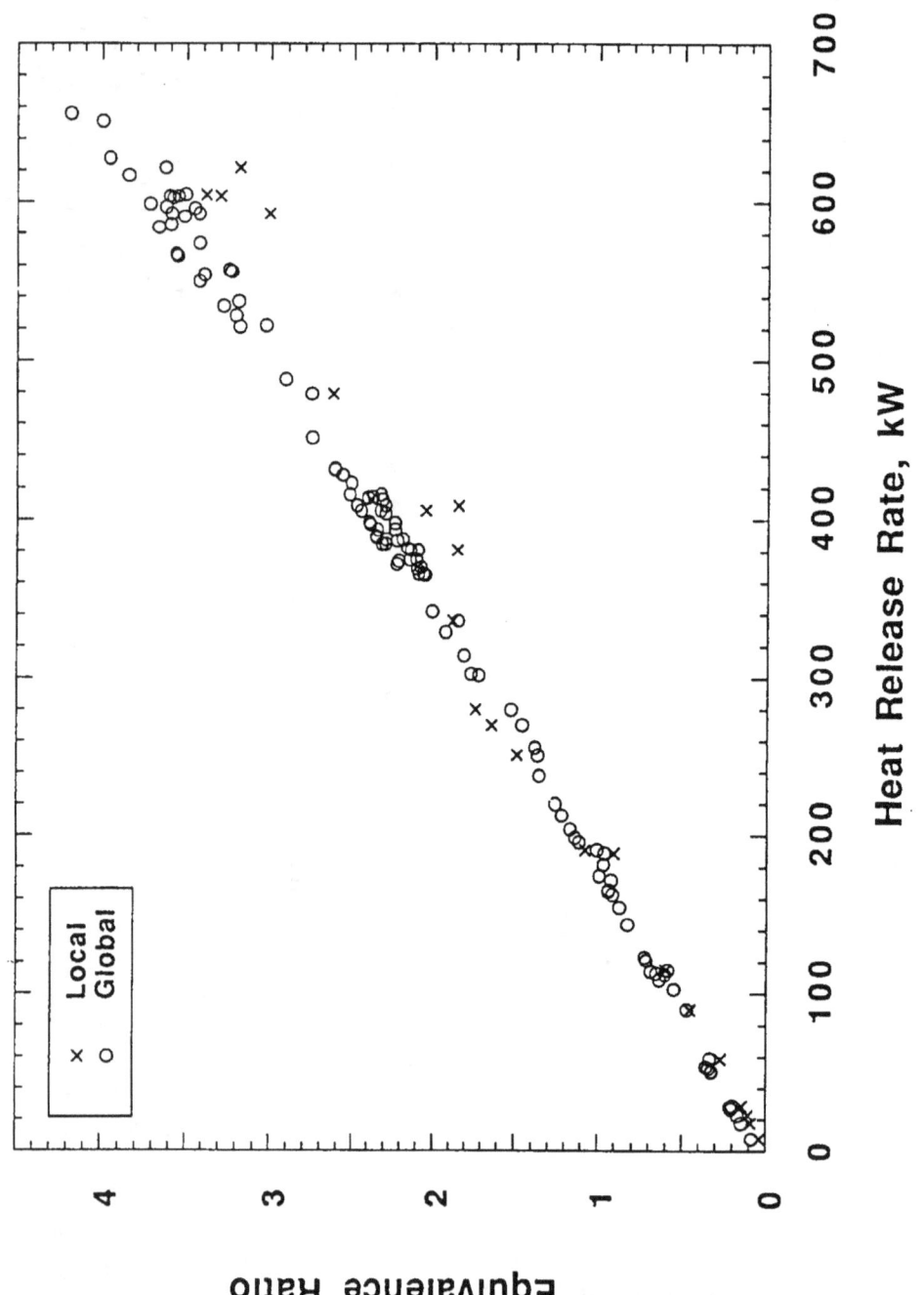

Figure 53.    Measured values of global equivalence ratio ($\phi_g$) calculated using the method of Janssens and Tran [52] and local equivalence ratio ($\phi_\ell$) using the $\phi$-meter (10 cm from front wall and ceiling, 29 cm from side wall) are plotted as a function of heat release rate for a series of natural gas fires in the RSE. Measurements were for pseudo-steady state burning.

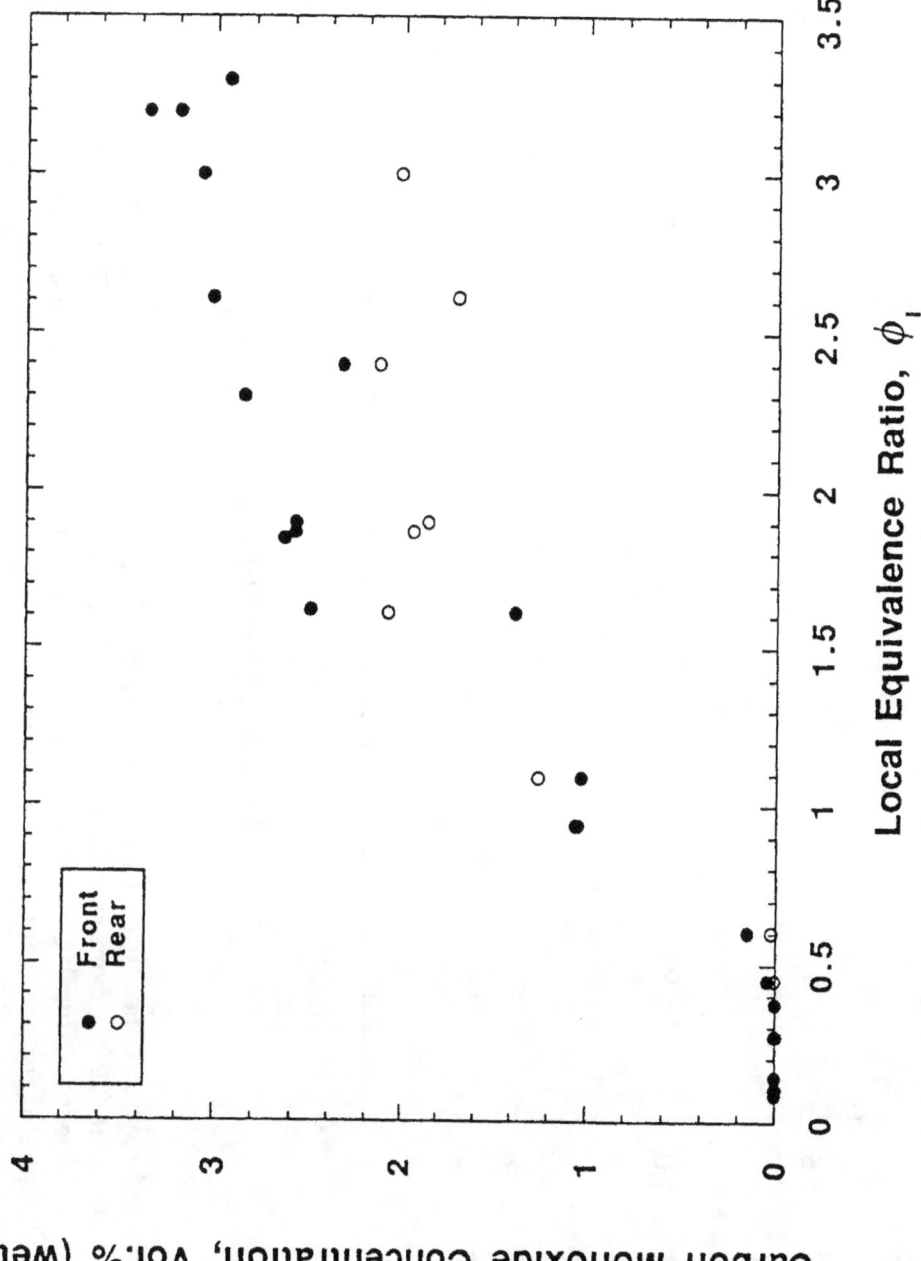

Figure 54.    Measured carbon monoxide concentrations for locations in the front (10 cm from front wall and ceiling, 29 cm from side wall) and rear (29 cm from rear and side walls, 10 cm from ceiling) of the RSE observed during pseudo-steady state burning of natural gas are plotted as a function of the local equivalence ratio $(\phi_\ell)$.  The $\phi$-meter measurement is for the front probe location.

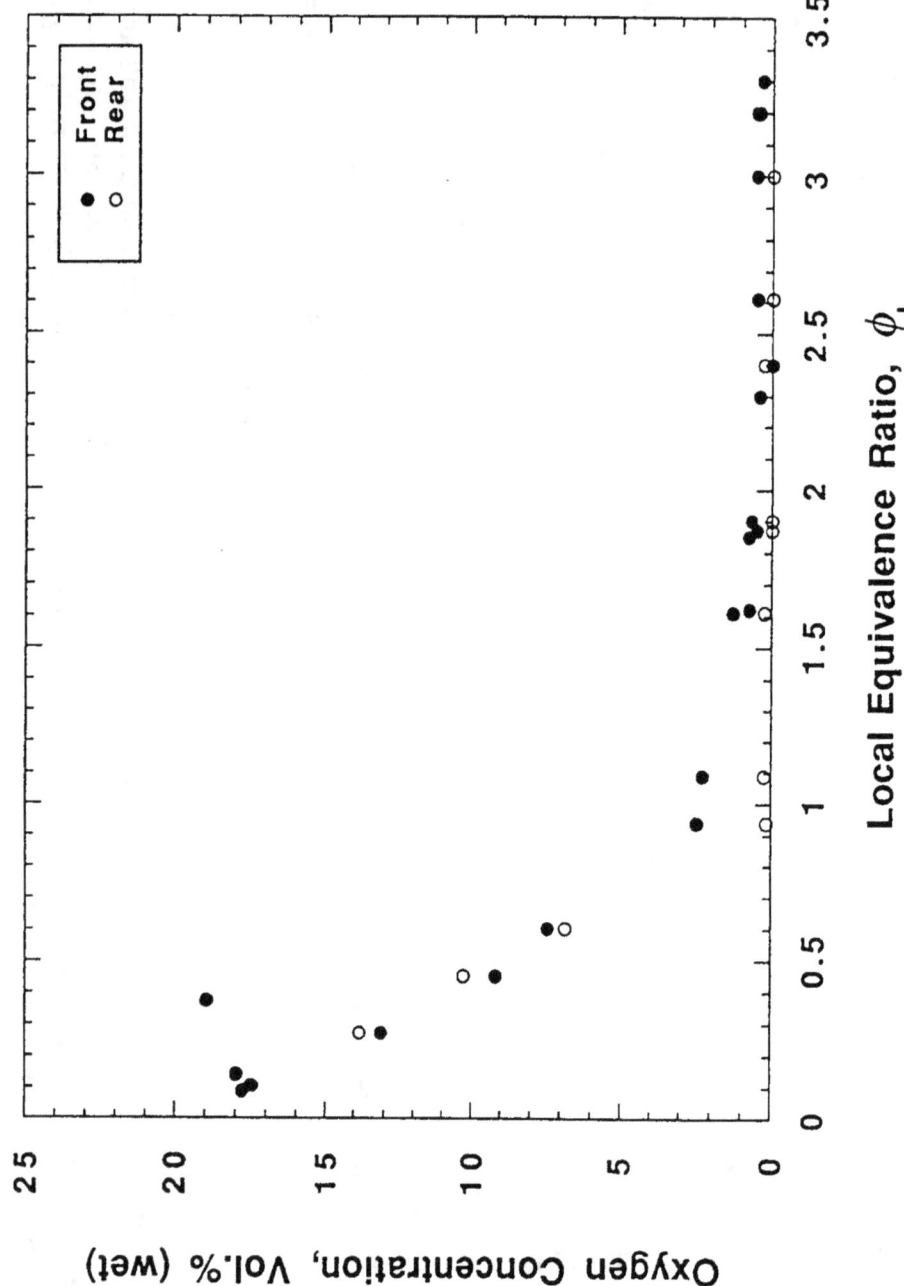

Figure 55. Measured oxygen concentrations for locations in the front (10 cm from front wall and ceiling, 29 cm from side wall) and rear (29 cm from rear and side walls, 10 cm from ceiling) of the RSE observed during pseudo-steady state burning of natural gas are plotted as a function of the local equivalence ratio ($\phi_\ell$). The $\phi$-meter measurement is for the front probe location.

CO concentration starts to increase at this value. It is clear that the $\phi$-meter measurements yield results which are consistent with the behaviors of the gas concentrations expected in the upper layer and with the global phi measurements. However, it must be remembered that the measurements are local in nature and, since values of $\phi_\ell$ seem to vary with position in the upper layer of the RSE, may not provide a true estimate for $\phi_g$.

It is of interest to compare CO, $CO_2$, and $O_2$ concentrations as functions of $\phi_g$ in the rear of the enclosure with those observed by Toner [22] in a hood experiment. Figures 56-58 compare the results for the two experiments. It is clear that the concentration behaviors observed in the rear of the enclosure are very similar to those measured by Toner in his hood.

2.      Experimental Findings for the RSE Equipped with a Narrow Door

A few fires were burned for which the doorway for the enclosure was narrowed to a width of 1 cm. For these conditions the ventilation is severely limited, and much smaller fires can generate rich upper-layer conditions than when the doorway is fully open. Due to the much lower HRR values employed, observed upper-layer temperatures were considerably cooler than for the fires discussed up to now. Visual inspection showed that these fires had a well defined layer interface and that mixing was less vigorous than in the larger fires. Based on these observations, it might be expected these fires would behave in a manner very much like the low-temperature hood experiments.

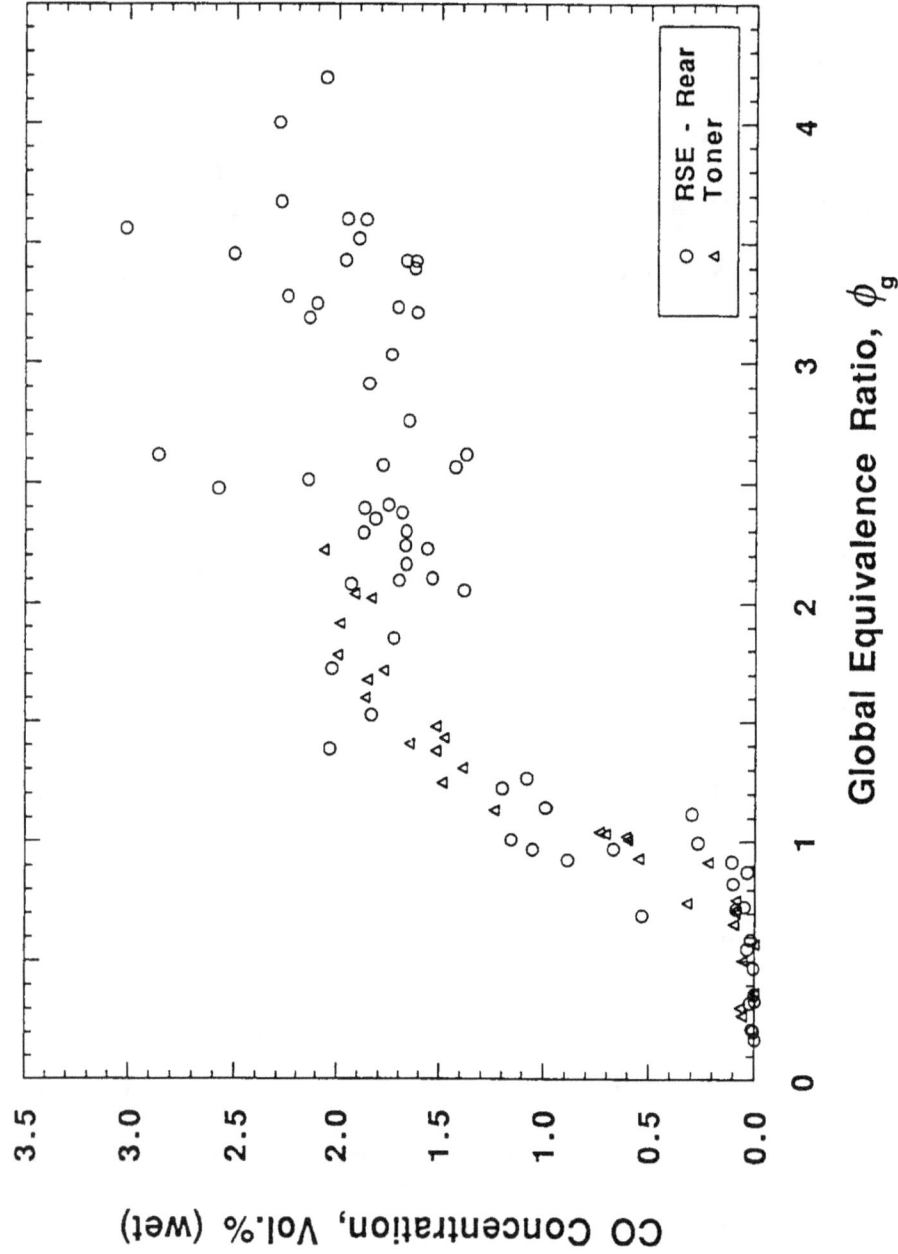

Figure 56. Values of carbon monoxide concentration observed for the upper layer in the rear (29 cm from rear and side walls, 10 cm from ceiling) of the RSE and in the hood experiments of Toner [22] are plotted as a function of the global equivalence ratios ($\phi_g$) observed in the two experiments. Natural gas was the fuel in both cases and measurements were recorded during pseudo-steady state burning.

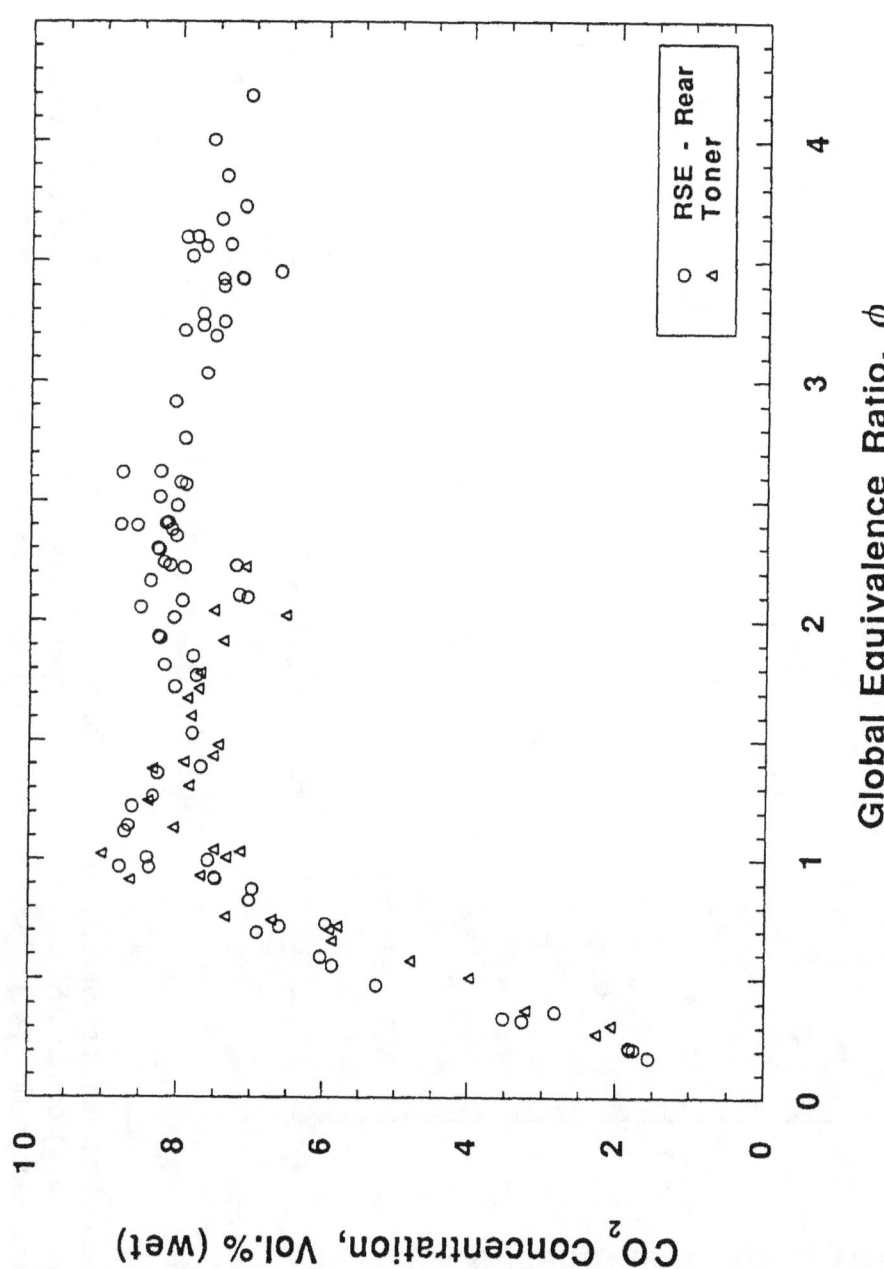

Figure 57.  Values of carbon dioxide concentration observed for the upper layer in the rear (29 cm from rear and side walls, 10 cm from ceiling) of the RSE and in the hood experiments of Toner [22] are plotted as a function of the global equivalence ratios ($\phi_g$) observed in the two experiments. Natural gas was the fuel in both cases and measurements were recorded during pseudo-steady state burning.

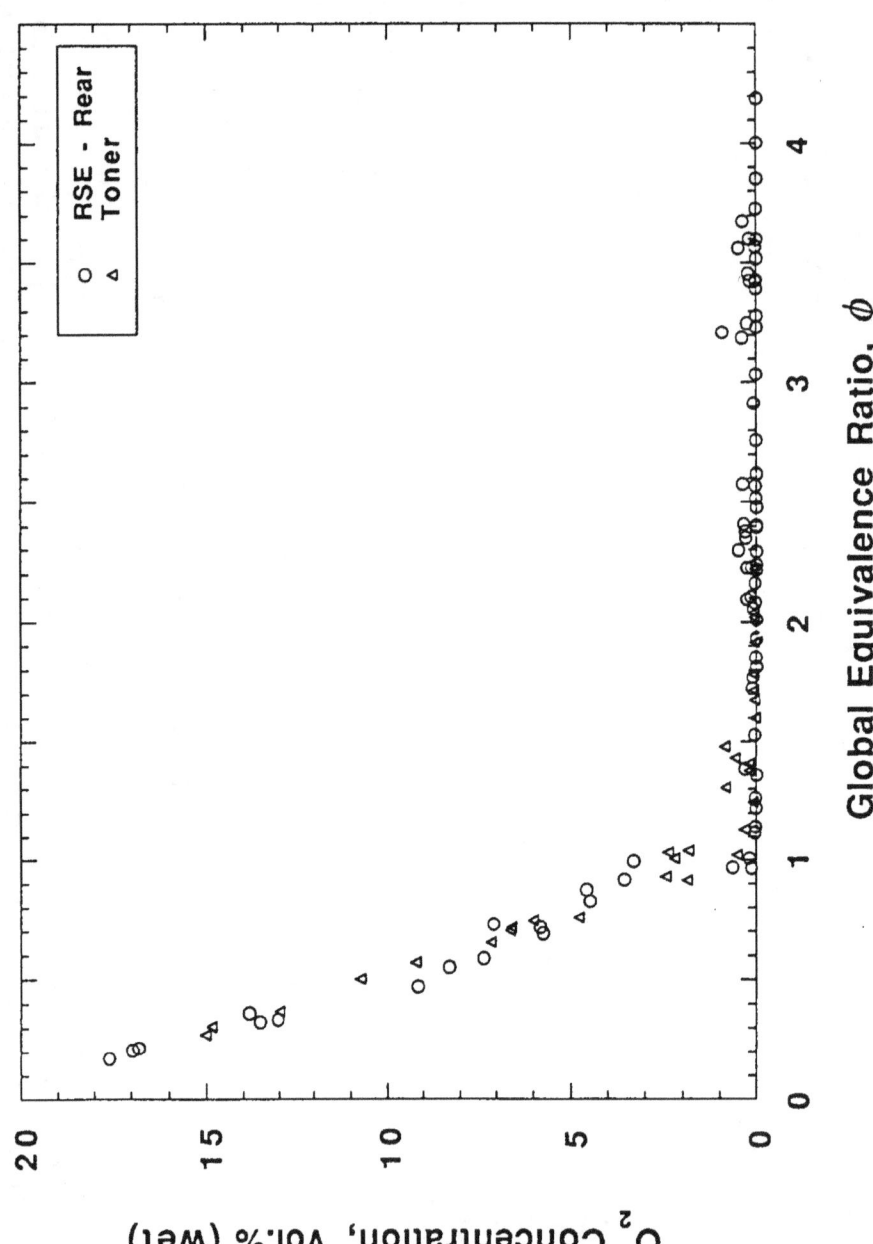

<u>Figure 58.</u>    Values of oxygen concentration observed for the upper layer in the rear (29 cm from rear and side walls, 10 cm from ceiling) of the RSE and in the hood experiments of Toner [22] are plotted as a function of the global equivalence ratios ($\phi_g$) observed in the two experiments. Natural gas was the fuel in both cases and measurements were recorded during pseudo-steady state burning.

108

No extensive concentration mappings of the upper layer were performed for these fires. Measurements were only recorded for two positions (both were 29 cm from the side wall and 10 cm from the ceiling, one was 10 cm and the second 29 cm from the front wall). The concentration measurements for these two positions were in good agreement and tracked each other well. Figure 59 shows the CO concentration time record for a 25 kW fire. It can be seen that the fire eventually reached a pseudo-steady state.

Temperature records were recorded in the front and rear of the enclosure. For all fires the maximum temperatures were less than 700 K, even though they did tend to rise slowly with time; presumably, as a result of heating of the enclosure walls. There was also a distinct gradient in the upper-layer temperature profiles with temperature decreasing with distance from the ceiling as can be seen in Figure 60 for a 25 kW fire.

During these small fires it was not possible to estimate $\phi_g$ in the manner used for the larger fires. However, the $\phi$-meter was used and values of $\phi_\ell$ are available for the sampling position closest to the front wall. Table 3 contains a tabulation of fire HRR; $\phi_\ell$; and CO, $CO_2$, and $O_2$ concentrations for the limited narrow doorway cases investigated, as well as results from Morehart's data [23] for hood experiments having similar $\phi_g$.

The results suggest that the CO concentrations are slightly lower in the upper layer for the enclosure fires, but, given the limited number of data points and the variation in measured values as the result of experimental uncertainty, the two sets of data are in excellent agreement. In particular, CO concentrations for the enclosure fires start to rise for $\phi_\ell > \approx 0.5$, and there are significant concentrations of $O_2$ measured in the upper layer for $\phi_\ell > 1$.

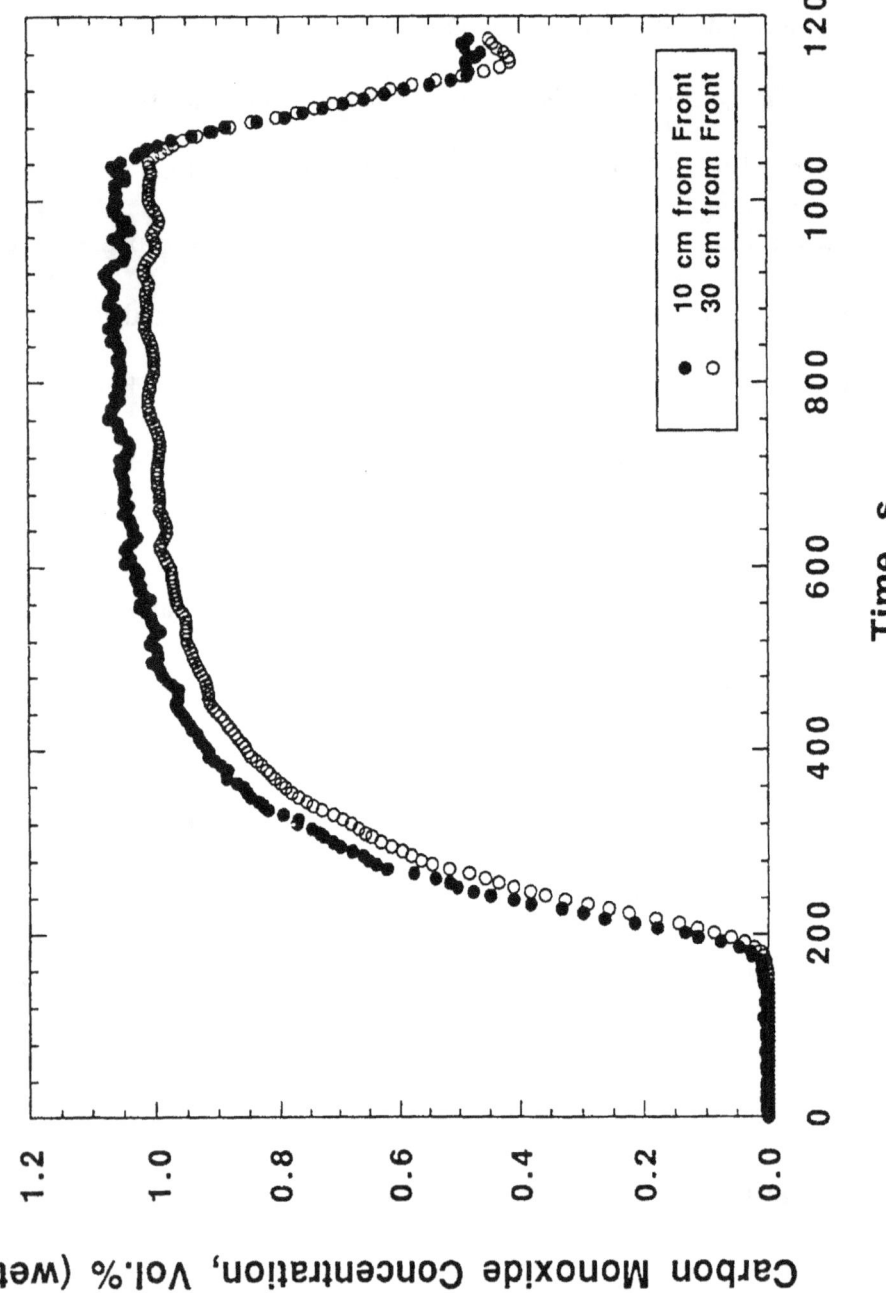

Figure 59.    Carbon monoxide concentrations measured in the upper layer of a 25 kW natural gas fire burning within the RSE having a narrow doorway (1 cm x 81 cm) are plotted as a function of time.  Measurement positions are both in the front (10 and 30 cm from the front wall, 10 cm from the ceiling, and 29 cm from the side wall).

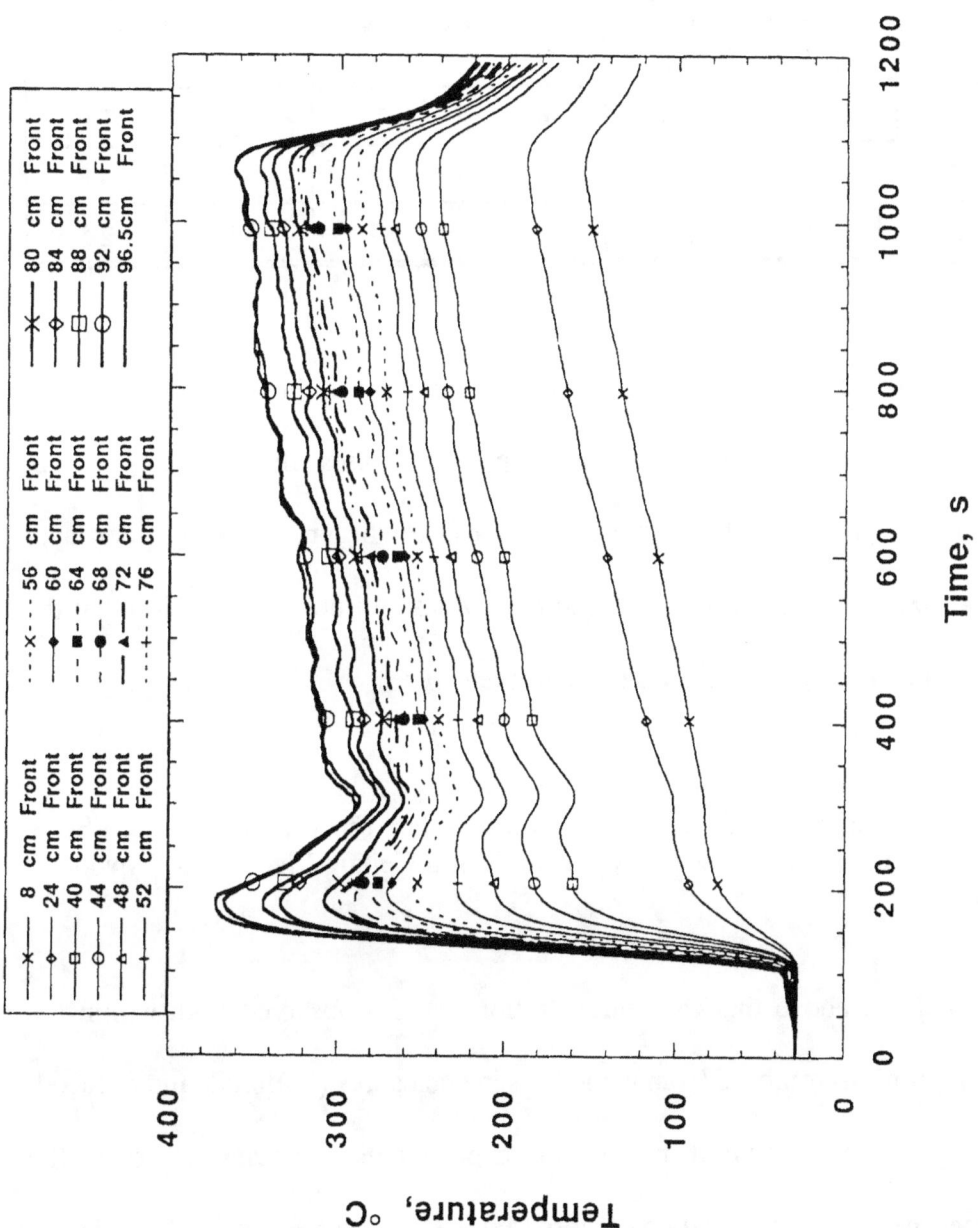

Figure 60. Temperature measurements from a vertical thermocouple tree located in the front (20 cm from front and side walls) of the RSE with a narrow doorway (1 cm x 81 cm) are shown as a function of time. A 25 kW natural gas fire was used.

Table 3.  Concentrations Observed in Low-Temperature Enclosure and Hood Fires

| HRR (kW) | $\phi_\ell$ | This Work | | | Hood Experiment (Morehart) | | |
|---|---|---|---|---|---|---|---|
| | | Concentration (Wet Vol. %) | | | Concentration (Wet Vol. %) | | |
| | | CO | $CO_2$ | $O_2$ | CO | $CO_2$ | $O_2$ |
| 7 | 0.45 | 0.003 | 4.7 | 10.0 | 0.00 | 5.1 | 9.6 |
| 10 | 0.68 | 0.23 | 6.5 | 6.5 | 0.35 | 6.2 | 7.0 |
| 15 | 1.2 | 0.8 | 7.1 | 3.9 | 1.3 | 6.9 | 3.8 |
| 25 | 1.5 | 1.0 | 7.3 | 2.1 | 1.5 | 6.9 | 2.5 |

These findings suggest that the results of the low-temperature hood experiments can be used to predict the concentrations of combustion gases for cases where the upper layer is well defined, no strong mixing of the upper and lower layers occurs, and the upper-layer temperature are < 700 K.  The close agreement also provides indirect collaboration that the prototype $\phi$-meter is providing accurate $\phi_\ell$ measurements.

3.     Experimental Findings for Natural-Gas Fires in a Reduced-Scale Standard Room with Upper Walls and Ceiling Lined with Plywood

It was pointed out above that the concentrations of CO observed both within the upper layer and for positions removed from wood fires in enclosures (both full- and reduced-scale) are often higher than measured in the hood experiments of Beyler [18] or in the enclosure fire experiments of Roby and coworkers [42],[43] where two well-defined layers were present.  Beyler has argued that high concentrations of CO are observed in the small-scale experiments as the result of sampling within fire gases [18]. However, this does not explain the high CO concentrations found at positions far removed from full-scale enclosure fires such as the Sharon fire test [14]. One must conclude that another mechanism for the formation of CO (besides the quenching of a fire plume in a rich upper layer located above the fuel) is responsible.

The enclosure experiments discussed earlier have shown that upper layers in fires having temperatures > 900 K contain very low concentrations of oxygen. It has also been argued that such oxygen-depleted layers do not react further until temperatures greater than 1350 K are achieved. This suggests that species produced by the pyrolysis of wood located within an upper layer of an enclosure fire could exit the enclosure unreacted. Since wood contains oxygen, it was hypothesized that the pyrolysis of wood in a hot upper layer could generate CO directly, which would provide an explanation for the unexpectedly high (based on the GER concept) CO levels observed in some enclosure fires with wood fuel. Some support for this hypothesis is found in pyrolysis investigations of wood under helium and nitrogen atmospheres [54],[55]. These studies have shown that, for temperatures typical of upper layers in fully developed fires, well over 50% of the mass of wood pyrolyzed generates gases which are primarily CO, $CO_2$, and $H_2O$ for high gas temperatures. The yield of $CO_2$ remains nearly constant as the pyrolysis gas temperature is increased, but the CO yield rapidly increases. For a temperature of 1100 K, Arpiainen and Lappi found that the pyrolysis of wood bark yielded 0.3 g CO per g wood pyrolyzed [55]. The production of CO was favored in a mole ratio of 5:1 over $CO_2$.

In order to test the above hypothesis, test fires were burned in the RSE for which the ceiling and upper walls (a 36 cm wide band around the inside of enclosure) were covered with 6.4 mm thick sheets of fir plywood [56]. Since it was desired to investigate conditions for which the upper-layer concentrations of $O_2$ were near zero and for which the upper-layer temperatures were high, natural-gas fuel flow rates were set to high nominal HRR values (550 kW and 600 kW). Upper-layer gases were sampled at the front and rear locations simultaneously, and the gas compositions were measured in the manner discussed earlier. Since the composition of the upper-layer gases is difficult to assess in the present experiments (the relative amounts of natural-gas combustion products and wood pyrolysis products are unknown), the measurements have not been corrected for water removal and are reported on a dry basis.

Figures 61-63 show time records for CO, $CO_2$, and $O_2$ measured in the front and rear of the enclosure for a 600 kW fire (based on natural-gas flow rate). Different stages in the fire are evident. During the period from $\approx$ 125 to 600 s the wood remained attached to the enclosure walls and ceiling; and, presumably, pyrolysis of the wood was taking place in the upper layer. The pyrolysis of the wood is evident from measurements of the HRR using the furniture calorimeter. Figure 64 shows that during the period when wood was pyrolyzing the HRR was significantly elevated above the nominal HRR for the natural gas alone, indicating that the wood pyrolysis products were burned on exiting the RSE.

At $\approx$ 600 s the wood was sufficiently structurally weakened that it could no longer support its own weight, and it fell in pieces to the floor of the enclosure. When the wood first fell it did not burst into flame, but instead appeared to be smoldering. After a short

114

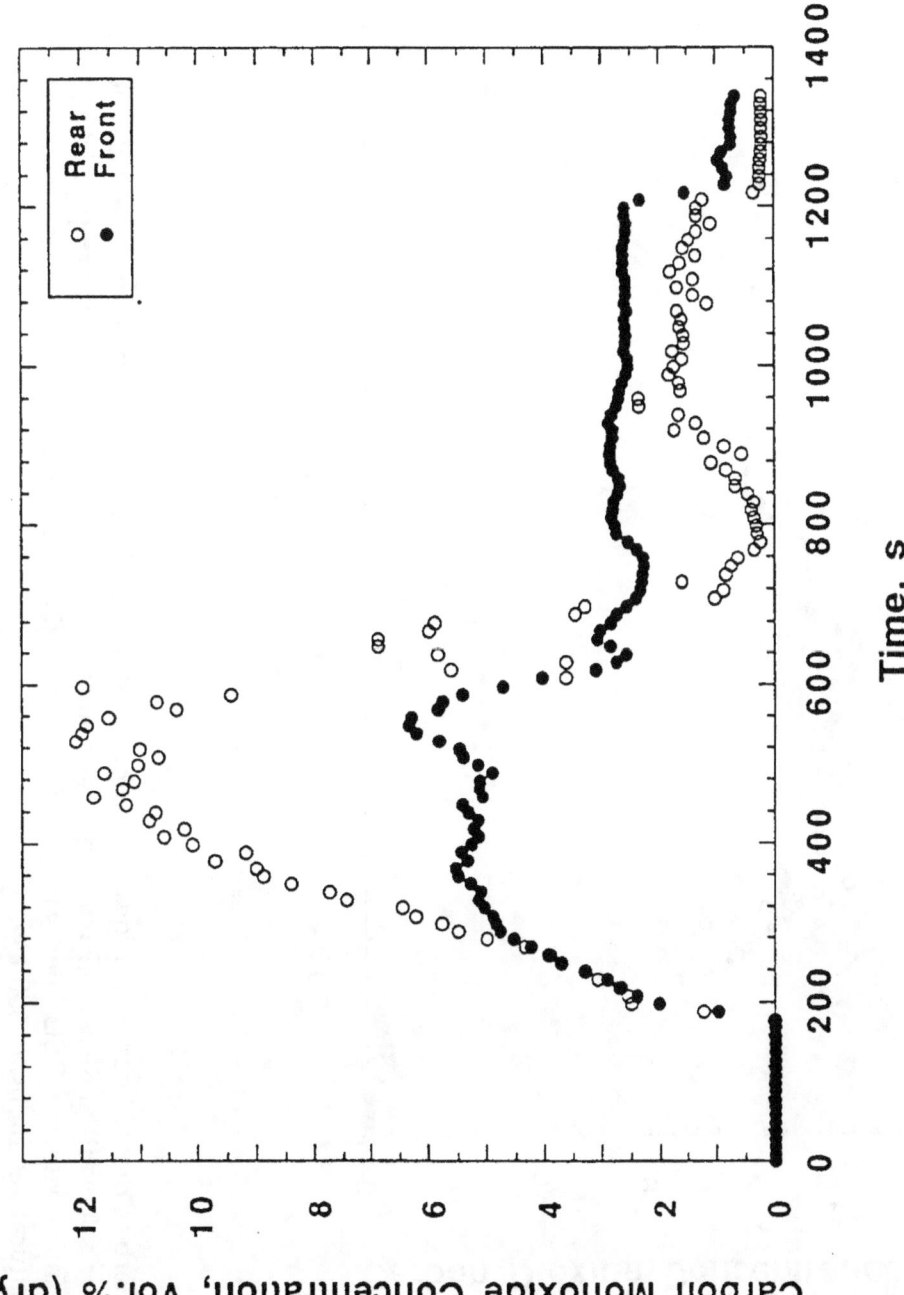

Figure 61. Carbon monoxide concentrations for probe locations in the front (10 cm from front wall and ceiling, 29 cm from the side wall) and rear (29 cm from rear and side walls, 10 cm from ceiling) are plotted as a function of time for a burn with the RSE lined with wood on the ceiling and upper walls. The nominal heat release rate for the natural gas fuel was 600 kW.

115

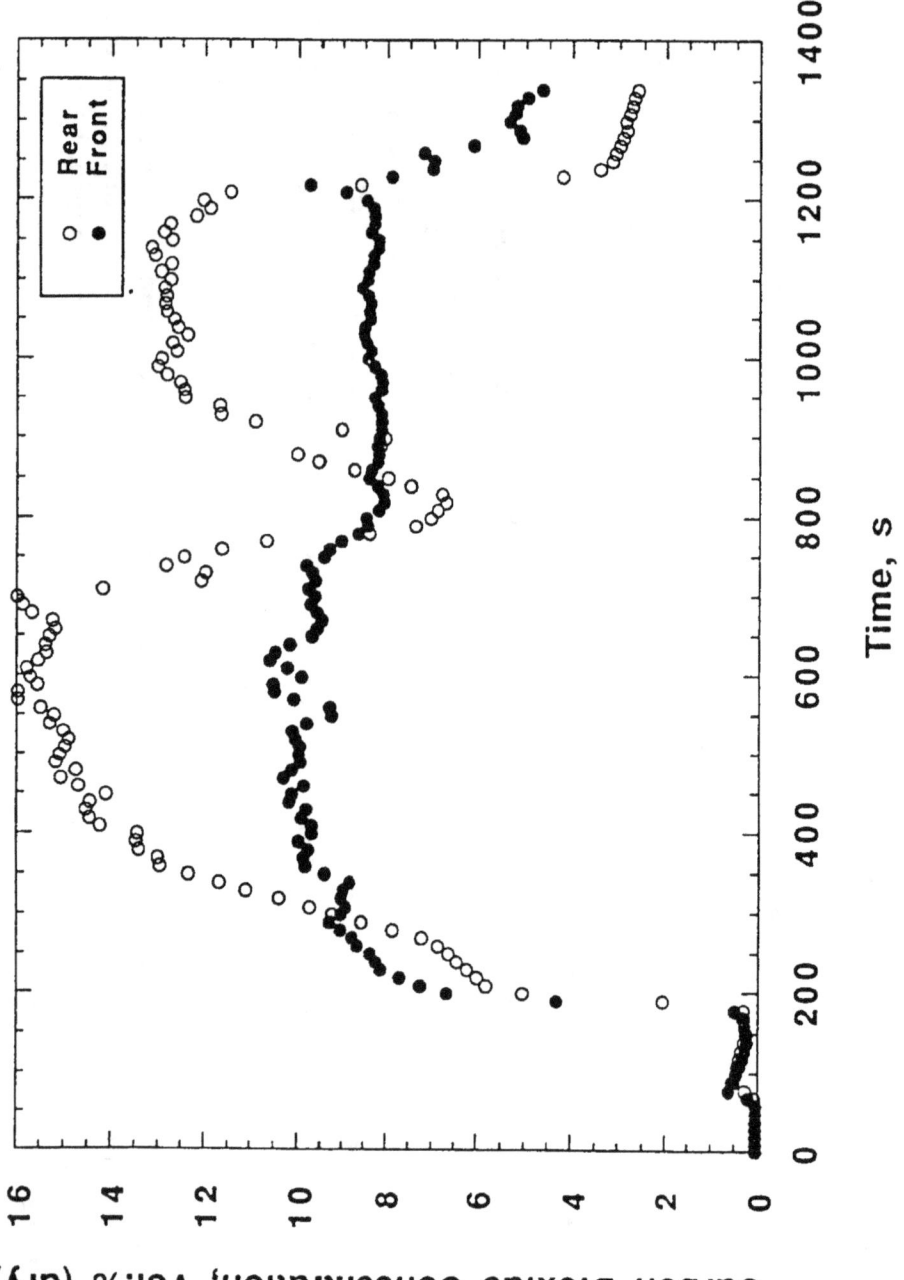

Figure 62. Carbon dioxide concentrations for probe locations in the front (10 cm from front wall and ceiling, 29 cm from the side wall) and rear (29 cm from rear and side walls, 10 cm from ceiling) are plotted as a function of time for a burn with the RSE lined with wood on the ceiling and upper walls. The nominal heat release rate for the natural gas fuel was 600 kW.

116

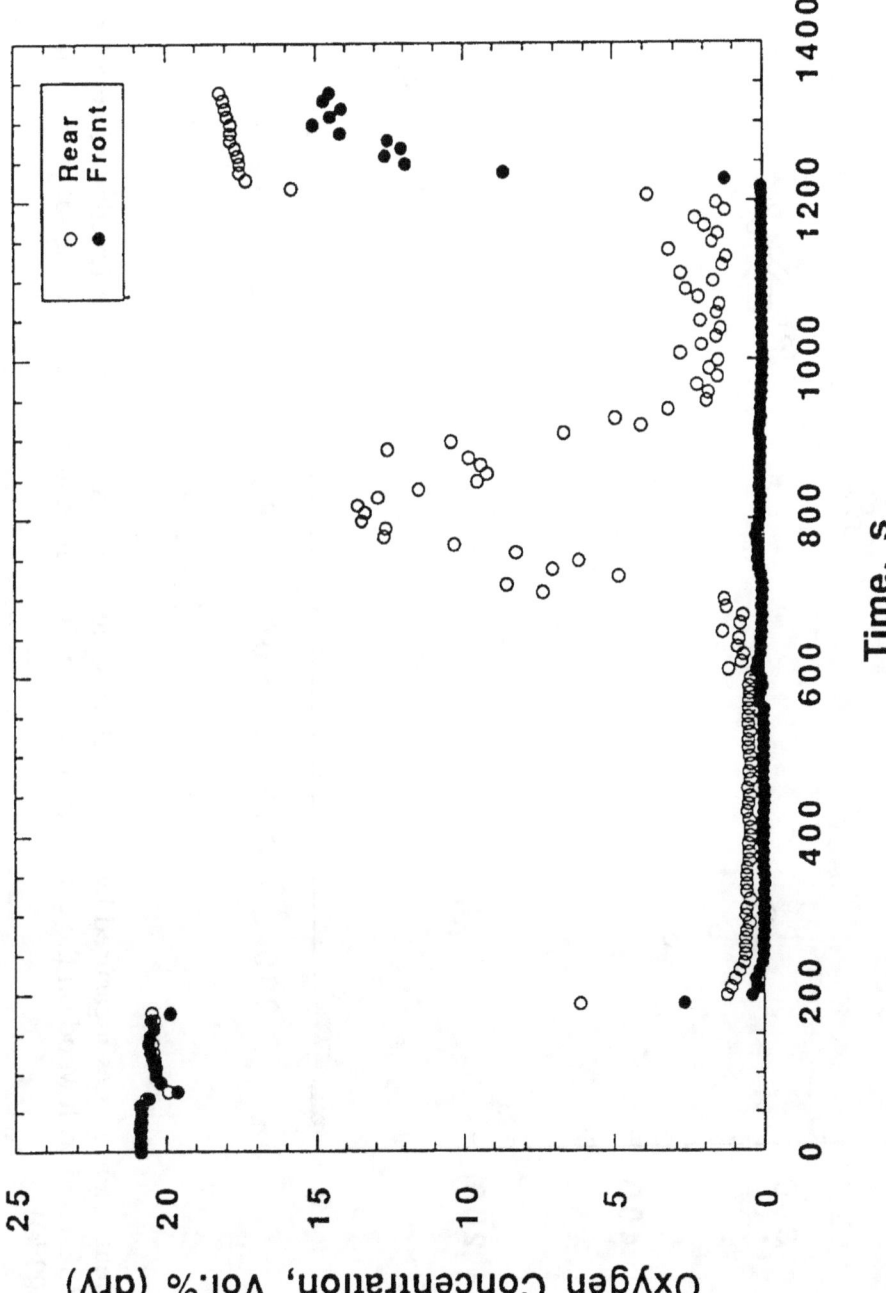

Figure 63.    Oxygen concentrations for probe locations in the front (10 cm from front wall and ceiling, 29 cm from the side wall) and rear (29 cm from rear and side walls, 10 cm from ceiling) are plotted as a function of time for a burn with the RSE lined with wood on the ceiling and upper walls.  The nominal heat release rate for the natural gas fuel was 600 kW.

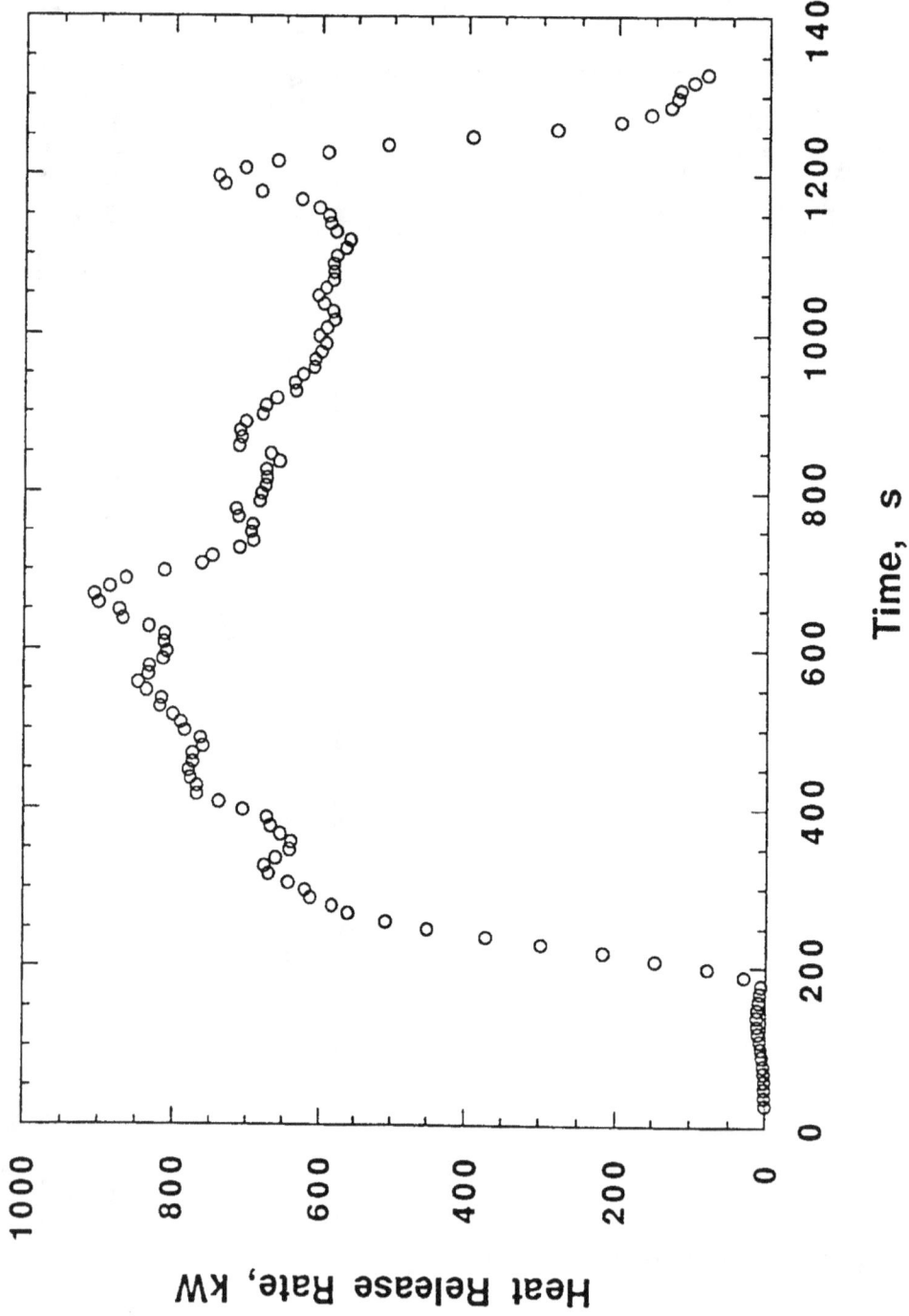

Figure 64. Heat release rates measured by the Furniture Calorimeter are plotted as a function of time for a burn with the RSE lined with wood on the ceiling and upper walls. The nominal heat release rate for the natural gas fuel was 600 kW.

period of time in the highly oxygenated lower layer the wood did begin to burn as evidenced by the attachment of flames.

The effect of the wood in the upper layer on the generation of CO was dramatic as can be seen by comparing figure 42 (for a 500 KW fire with no wood) and figure 61. In the rear of the enclosure, dry CO concentrations reached levels > 12% (note that the CO concentration may actually have been higher since this is close to the maximum range of the CO analyzer). Increases in CO concentration for the front are not as dramatic, but levels approaching 6% were observed. This should be compared with measurements of roughly 3.0% (wet) in the front and 2.0% (wet) in the rear for fires with natural-gas fuel alone.

$CO_2$ concentrations were also increased in both the front and rear of the enclosure compared to the case with no wood (compare the levels in fig. 62 with those for steady-state burning in fig. 46). However, the relative increases were much smaller than for CO, suggesting that considerably more CO than $CO_2$ was produced as predicted by the pyrolysis studies discussed above [54],[55]. Figure 63 shows that oxygen concentrations were below 1% for both locations in the enclosure, even though the residual level did seem to be slightly higher in the rear.

The concentration behaviors changed dramatically after the wood collapsed. In the front of the enclosure, CO levels decreased to those typical of a natural-gas only fire. The rear CO concentration behavior was more complicated. First the CO levels dropped to very low values (< 1%) and then recovered to a value of ≈ 1.8%, which is slightly less than found for the earlier fires. On the other hand, the $CO_2$ level first fell rapidly, but then recovered to ≈ 13% which is higher than for natural gas alone. Figure 63 shows that when the wood fell there was a large increase in $O_2$ concentration suggesting that air was mixed into the rear upper layer by the falling wood. Eventually, the $O_2$ level decreased to a residual level which was > 2%. These observations suggest that conditions in the rear of the enclosure are different when only natural gas is burned. Perhaps the presence of the wood burning on the floor is responsible.

Figures 65 and 66 show temperature measurements from thermocouple trees located in the rear and front of the enclosure, respectively. Close inspection of figure 65 shows some very interesting effects. First note that the thermocouple for the highest position (96.5 cm from the floor) has recorded relatively low temperatures during the pyrolysis of the wood, rising from 300 K to 600 K. Most of the remainder of the upper layer seems to have uniform temperatures $\approx$ 150 K higher. The highest temperatures seem to be near the bottom of the layer. The measured temperatures are considerably cooler than observed in the rear of the enclosure for natural-gas only fires. These observations suggest that the pyrolysis of the wood is taking place at a surface temperature below the upper-layer gas temperature and that the pyrolysis and subsequent heating of the gases generated requires enthalpy extraction from the layer. Following the collapse of the wood, the temperature quickly increases to levels typical of the natural-gas only fires. Note that the thermocouple near the floor also recorded high temperatures during this period due to the presence of burning wood on the floor.

The effects of the wood pyrolysis on the temperatures in the front of the enclosure (fig. 66) are not nearly as dramatic. The highest location (96.5 cm) does have a temperature significantly lower than other locations in the upper layer, and it also increases during the

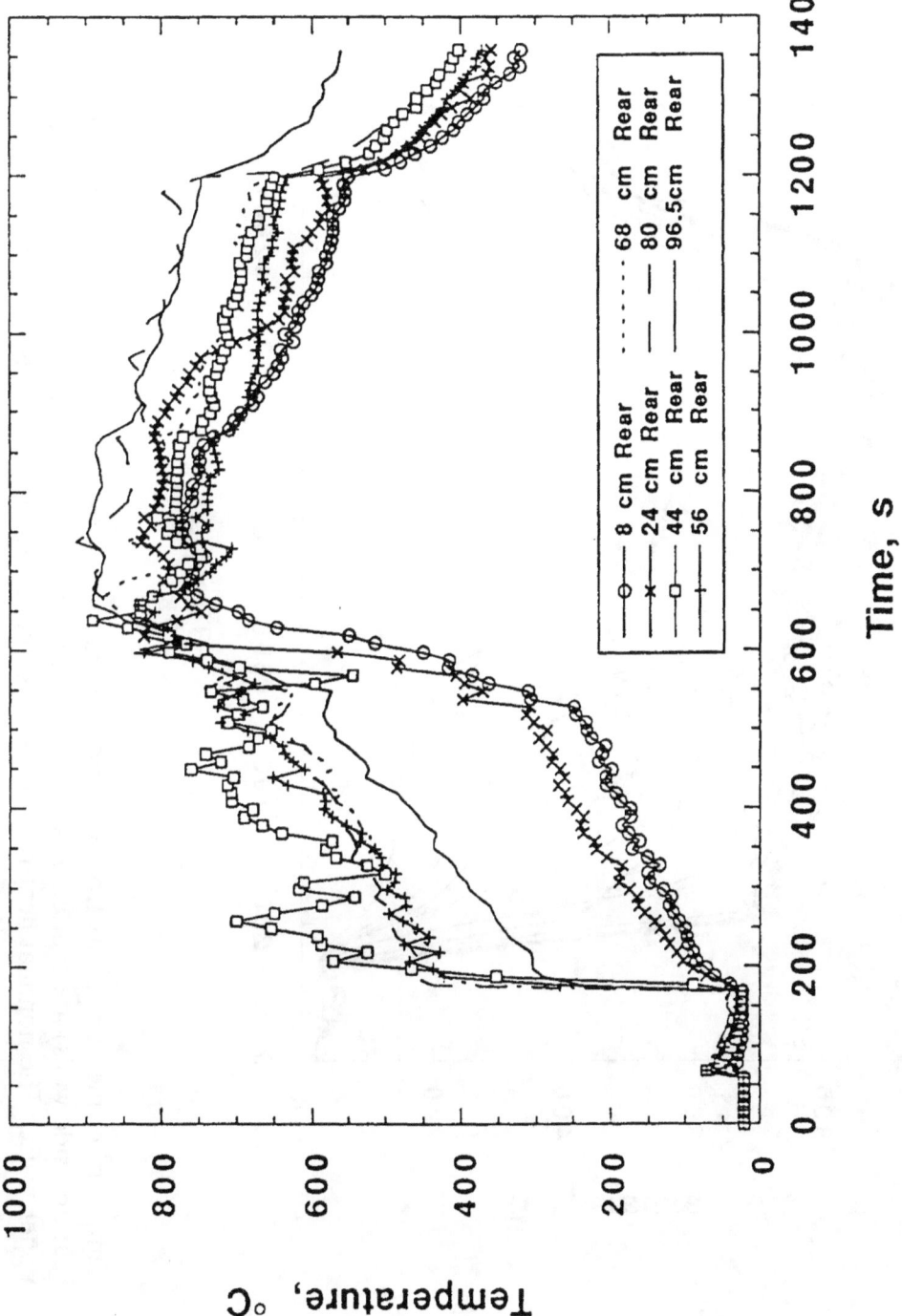

Figure 65.  Temperature measurements from a vertical thermocouple tree located in the rear of the RSE (20 cm from the front and side walls) are plotted as a function of time for a burn with the RSE lined with wood on the ceiling and upper walls.  The nominal heat release rate for the natural gas fuel was 600 kW.

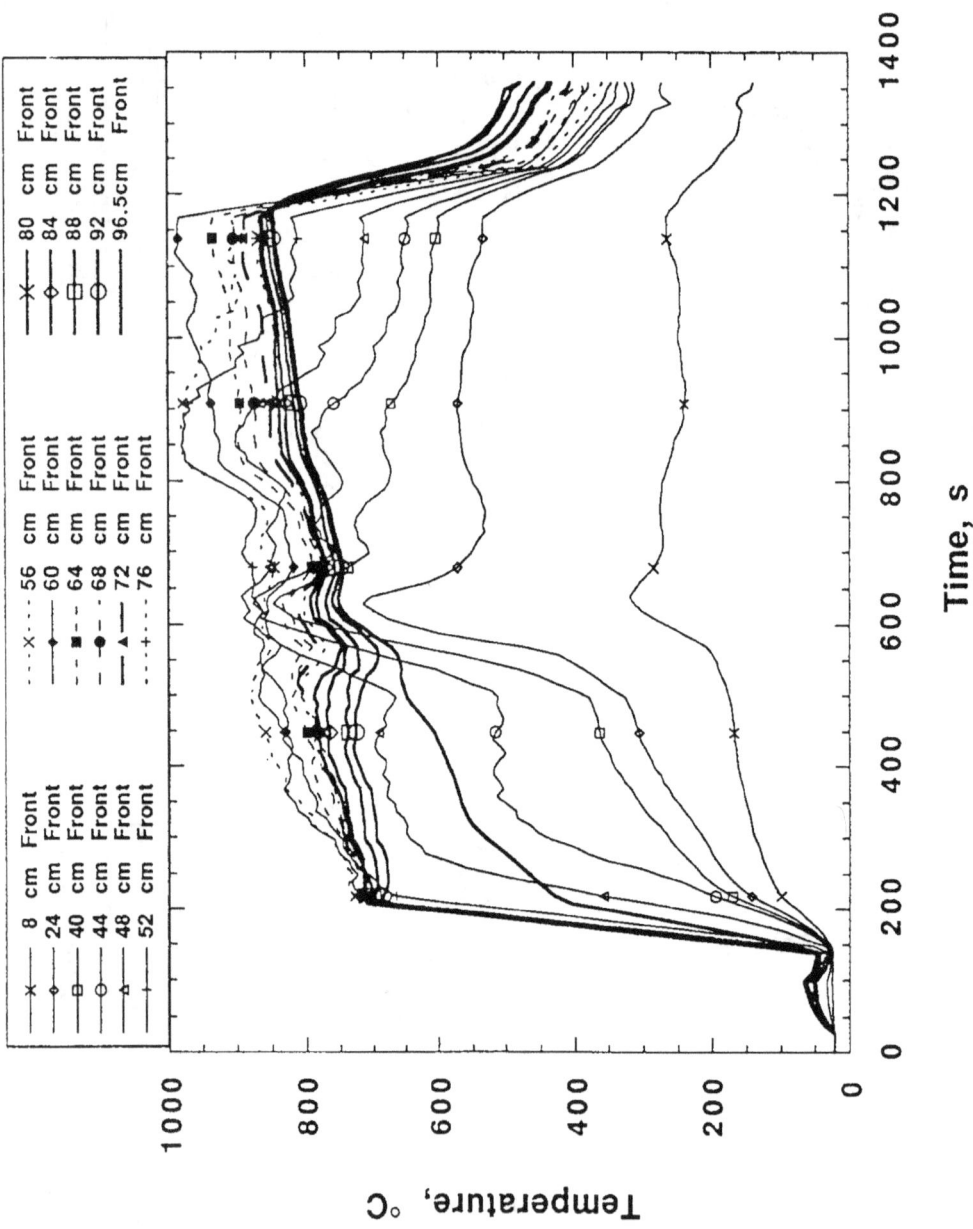

Figure 66. Temperature measurements from a vertical thermocouple tree located in the front of the RSE (20 cm from the rear and side walls) are plotted as a function of time for a burn with the RSE lined with wood on the ceiling and upper walls. The nominal heat release rate for the natural gas fuel was 600 kW.

period of wood pyrolysis. Again, this is taken to be indicative of the wood pyrolysis and heating of pyrolytic gases requiring a net input of heat. The other thermocouple measurements are found to reach relatively constant temperature levels fairly quickly, suggesting rapid mixing. Following the collapse of the wood there are small increases in temperature in this region of the layer.

The findings discussed in this section demonstrate that the presence of wood in an oxygen-depleted upper layer of an enclosure fire can lead to the generation of significantly higher concentrations of CO as suggested by wood pyrolysis investigations [54],[55]. The results indicate that, for these conditions, the generation of gaseous fuel by the pyrolysis of wood is an endothermic process. The results of the wood pyrolysis for locations in the front and rear of the enclosure are very different. A possible explanation for these differences is discussed in Section V.C.5.

4.      Field Modeling Study of Flow Fields and Mixing in the RSE

It is clear from the above discussion that the upper layers in RSE fires are far from homogeneous and that reactions which generate CO must be taking place within the upper layer. Since it has not been possible to make experimental measurements of velocity fields or mixing behavior within the RSE, a computational effort has been initiated employing a three-dimensional k-$\epsilon$ model [57].

The commercial code used for these calculations is *FLOW3D* [58] which has been developed by Harwell Laboratories, Oxfordshire, UK. Earlier work at NIST has demonstrated that this model provides reasonable predictions of full-scale room fire behavior [59].

The calculations were run using a variable-size, rectangular 16 x 36 x 18 grid system superimposed on one half of the enclosure. The symmetry of the enclosure and burner position justified this approach. The burner was modeled by assuming fuel was released at four rectangular grids covering an area of 12 cm x 6 cm (again representing half of the full burner by symmetry). Flows near the wall were treated using standard wall functions [58].

The walls were assumed to remain at a constant temperature and heat losses were approximated by assuming that a fixed fraction of the heat release was loss by conduction and/or radiation. The remainder of the heat release produced the convective flows within the enclosure. Radiation from the hot gases and walls was ignored. One effect of neglecting radiation was that the walls and floor of the enclosure not in contact with the combustion gases were not heated. This approximation is not consistent with the experiments. Kumar et al., have investigated the effects of neglecting radiation on calculated fire behavior using a field model [60]. They concluded that the changes in calculated behaviors can be quite significant.

Combustion was simulated using the eddy breakup model [61], where the degree of combustion is controlled solely by the rate of mixing of fuel and oxidizer. Burning was assumed to be complete and to occur instantaneously when fuel and air were mixed. Note that this model does not allow modeling of CO formation or other products of incomplete combustion.

The experimental case modeled was that of a 400 kW fire centered in the RSE. As a test of the model, the predicted temperature distributions were compared with the experimental results for locations near the front (fig. 67) and rear (fig. 68) of the enclosure. The agreement for the front position is quite good even though it is evident that some heating of the lower layer does occur in the experiment. In the rear of the enclosure the agreement of experiment and calculation is much poorer. The model does predict a higher temperature within the upper layer for the front of the enclosure, but significantly overestimates the upper-layer temperature in the rear of the room. Significant heating of the lower layer is observed experimentally in the rear of the enclosure.

Figure 69 shows the calculated velocity profiles within a vertical plane oriented parallel to the long side of the RSE and located along the centerline. Several features of the flow field can be seen. The plume formed in the center of the enclosure by the fire is obvious. The gases in the fire plume rise, hit the ceiling, and form a ceiling jet moving away from the fire plume. The two-layer nature of the fire is revealed by changes in the direction of the flows. In particular, in the front of the enclosure, gases in the lower layer are moving towards the fire, while gases in the upper layer are moving towards the doorway. In the rear of the enclosure there is an apparent recirculation zone within the upper layer. The hot gases which exit the doorway are rapidly accelerated by buoyancy forces.

Figure 70 shows velocity components for a plane parallel to that in figure 69, but located 24 cm from the centerline of the enclosure. Many of the features observed on the centerline are present here as well. The recirculation zone in the rear of the upper layer is more clearly defined. It also appears as if this recirculation zone is entraining air directly

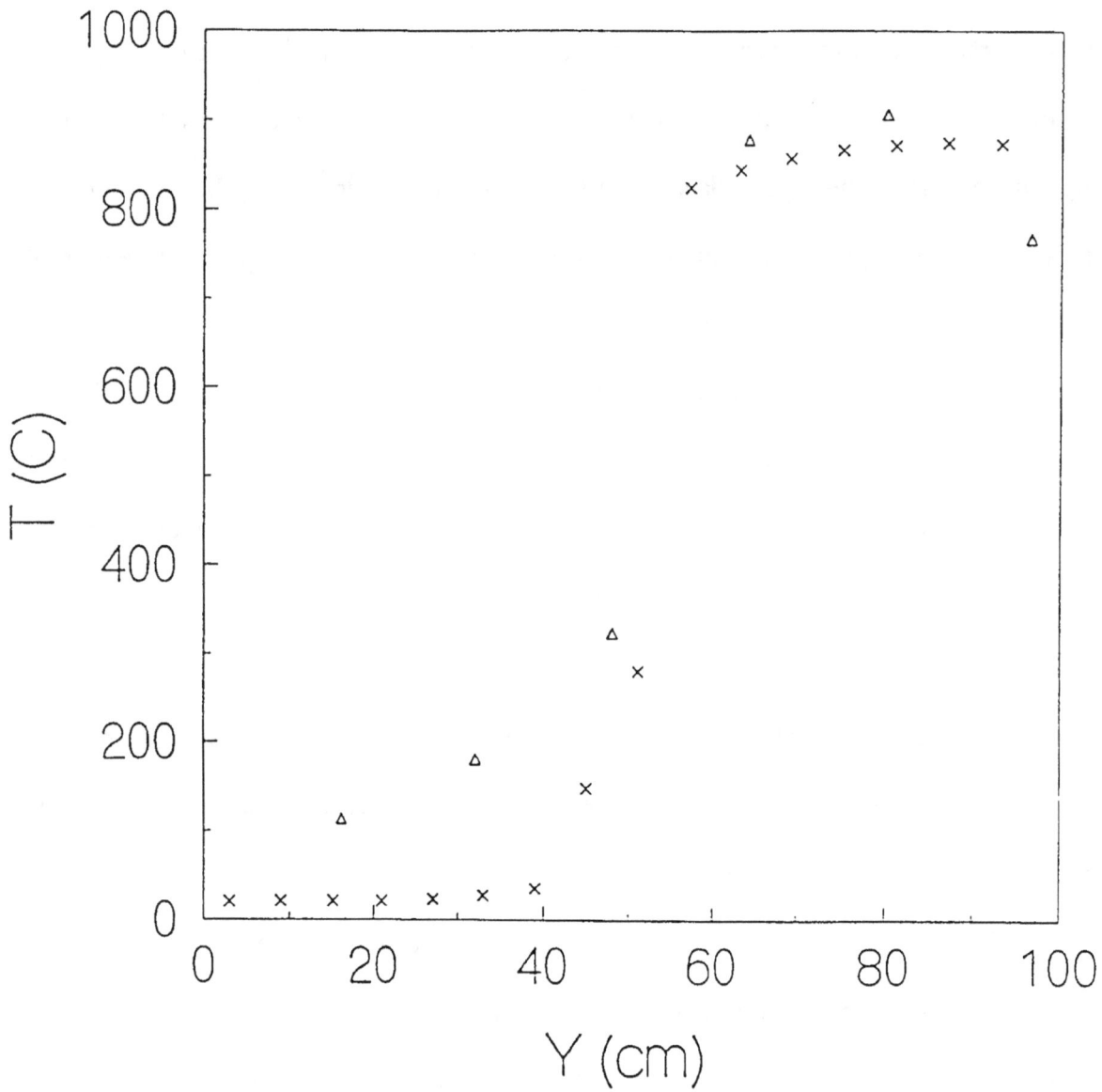

Figure 67.   Comparison of vertical temperature profiles measured (△) and calculated (x) using FLOW3D for a location 20 cm from the front and side walls of the RSE for a 400 kW natural gas fire.

126

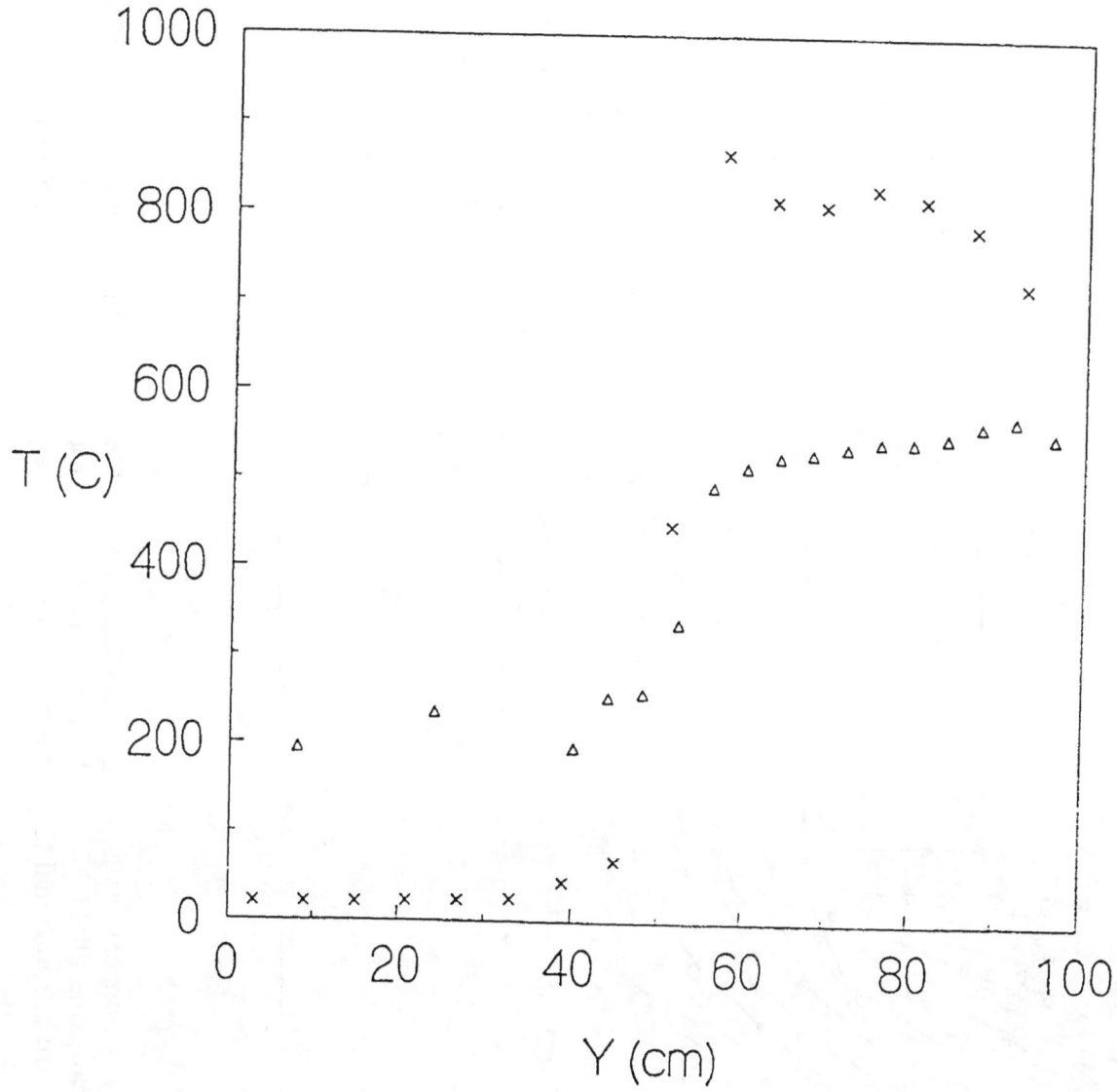

Figure 68.    Comparison of vertical temperature profiles measured (△) and calculated (x) using FLOW3D for a location 20 cm from the rear and side walls of the RSE for a 400 kW natural gas fire.

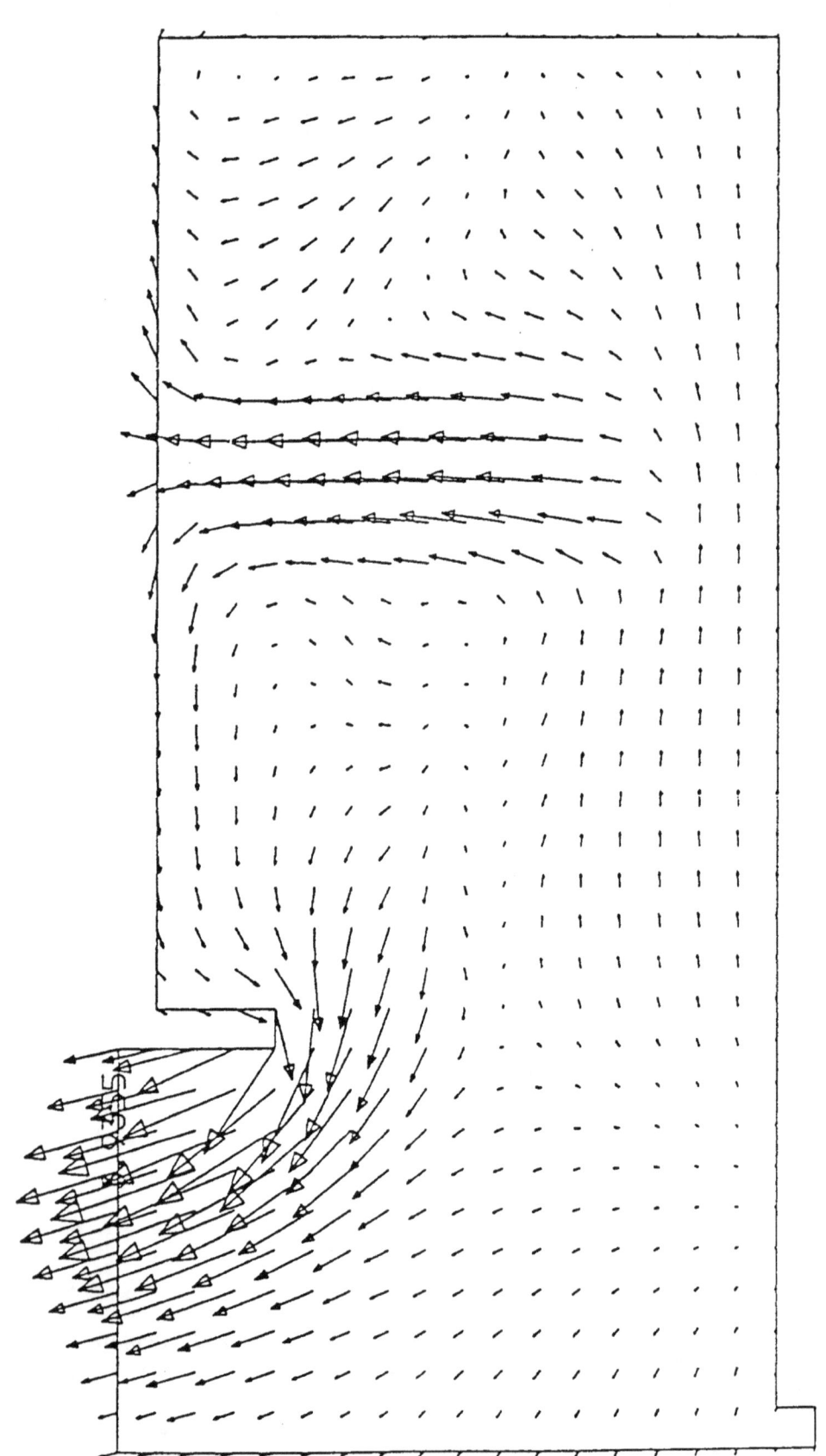

Figure 69. Velocity vectors calculated using FLOW3D for a 400 kW fire positioned in the center of the RSE are shown for a vertical plane aligned along the centerline of the enclosure parallel to the long direction. The lengths of the vectors are proportional to the velocity with the longest vector representing 3.94 m/s.

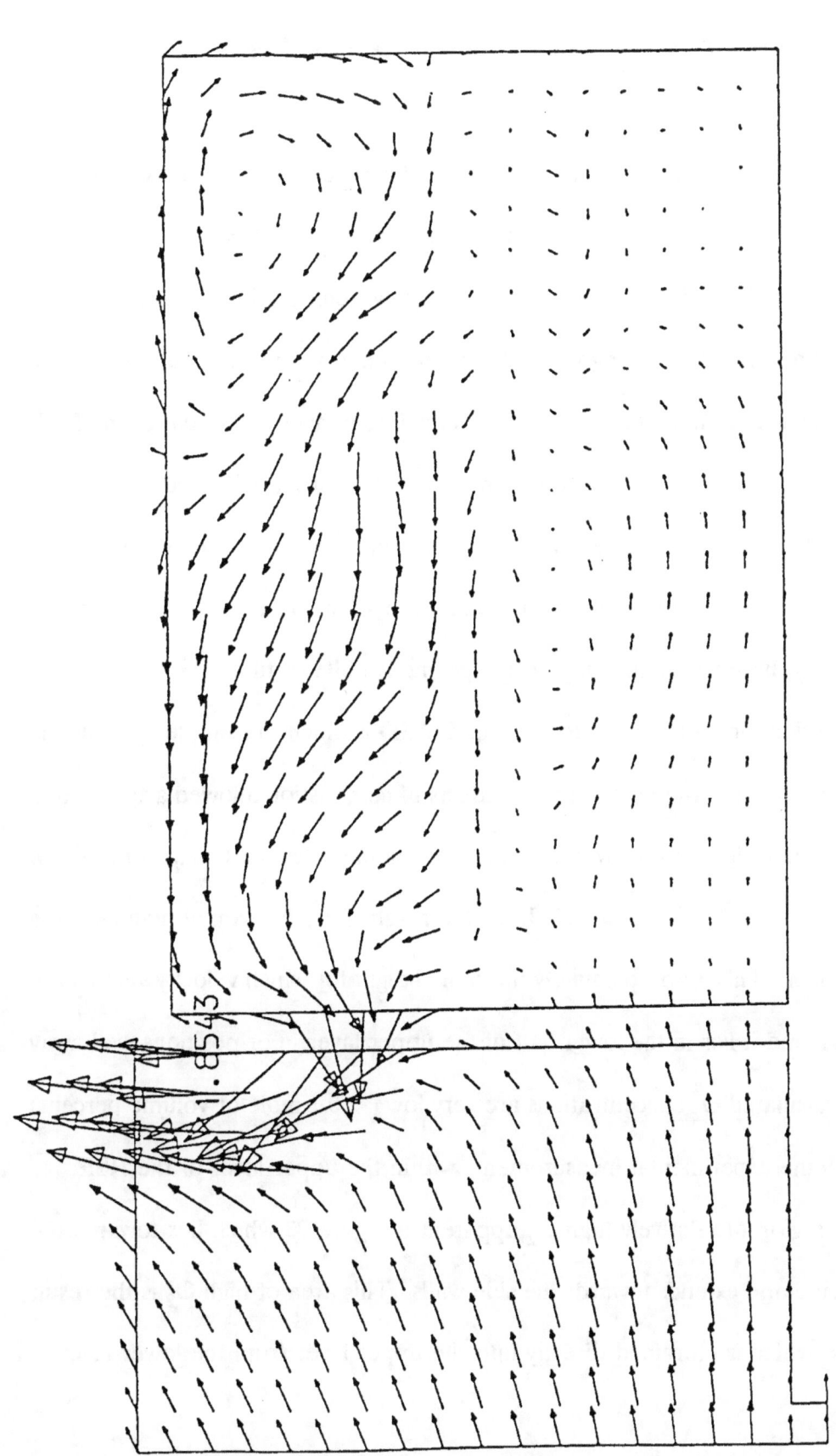

1.8543

<u>Figure 70.</u>    Velocity vectors calculated using FLOW3D for a 400 kW fire positioned in the center of the RSE are shown for a vertical plane positioned 24 cm from the centerline of the enclosure parallel to the long direction. The lengths of the vectors are proportional to the velocity with the longest vector representing 1.85 m/s.

from the lower layer and ultimately transporting it into the upper layer in the front of the RSE.

The velocity vectors calculated for a horizontal plane located 60 cm above the floor of the RSE are shown in figure 71. Only half of the enclosure is represented. This vertical position is in the lower half of the upper layer. The flow behavior in this plane is quite complex. There appears to be a strong vortical motion around the fire plume. One of the results of this motion is to transport air, which enters the upper layer in the rear of the RSE, to the front of the upper layer. Gases in the front of the RSE are accelerated towards the center by the need to exit through the doorway. Calculated velocity vectors in a plane located very near the ceiling (90 cm above the floor) are shown in figure 72. The radial ceiling jet formed by the plume impinging on the ceiling is quite distinct.

The combustion model incorporated into FLOW3D is much too simple to allow the prediction of CO formation. In fact, the only products of combustion allowed are $CO_2$ and $H_2O$. However, some insights are provided by considering the calculated concentrations of $O_2$ for locations within the RSE. Figure 73 shows the results for a horizontal plane located 63 cm above the floor. This is approximately the same height for which velocity vectors are shown in figure 71, and it lies in the lower half of the upper layer. For positions well away from the plume, calculated $O_2$ concentrations are very low ($< 0.1$ mole or volume percent) in agreement with the experimental measurements within the upper layer of the RSE.

There is a region of relatively high $O_2$ apparent in figure 73 which lies towards the rear of the enclosure and extends towards the side walls. This area of high $O_2$ is the result of transport of $O_2$ which is entrained directly into the upper layer from the lower layer in

130

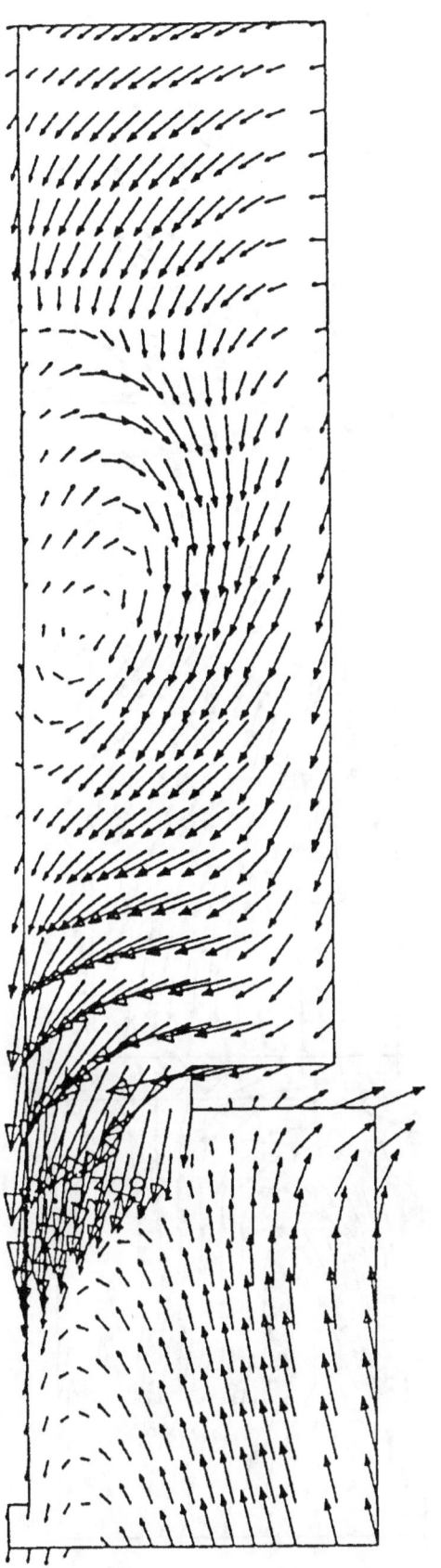

Figure 71. Velocity vectors calculated using FLOW3D for a 400 kW fire positioned in the center of the RSE are shown for a horizontal plane positioned 60 cm above the floor of the enclosure. Due to the symmetry of the problem, only half of the enclosure is represented. The lengths of the vectors are proportional to the velocity with the longest vector representing 1.54 m/s.

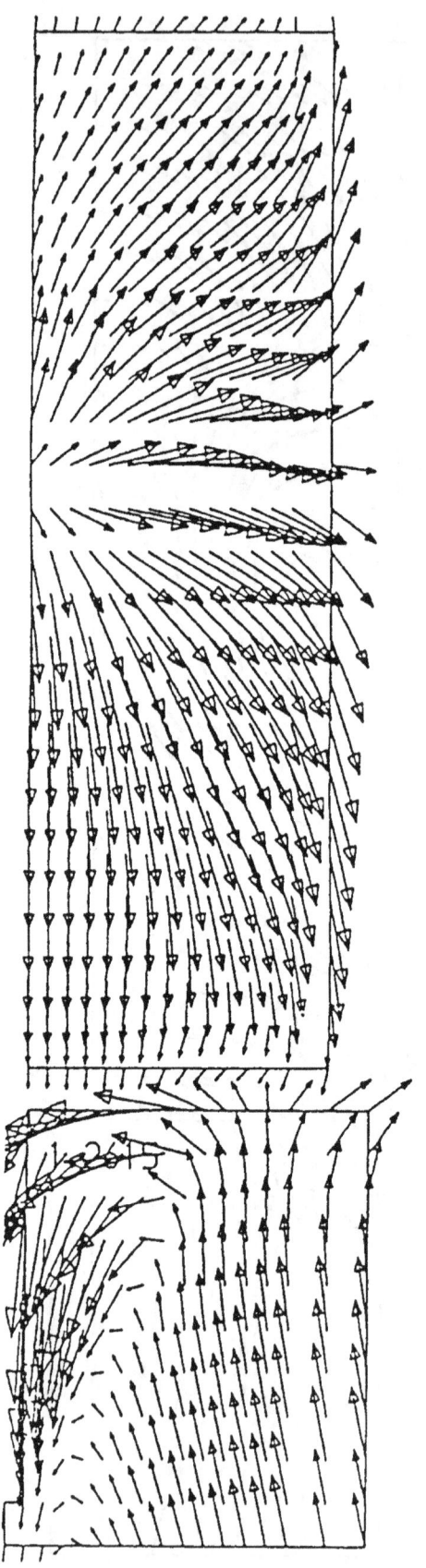

<u>Figure 72.</u>  Velocity vectors calculated using FLOW3D for a 400 kW fire positioned in the center of the RSE are shown for a horizontal plane positioned 90 cm above the floor of the enclosure.  Due to the symmetry of the problem, only half of the enclosure is represented.  The lengths of the vectors are proportional to the velocity with the longest vector representing 1.26 m/s.

# Oxygen Mole Fraction

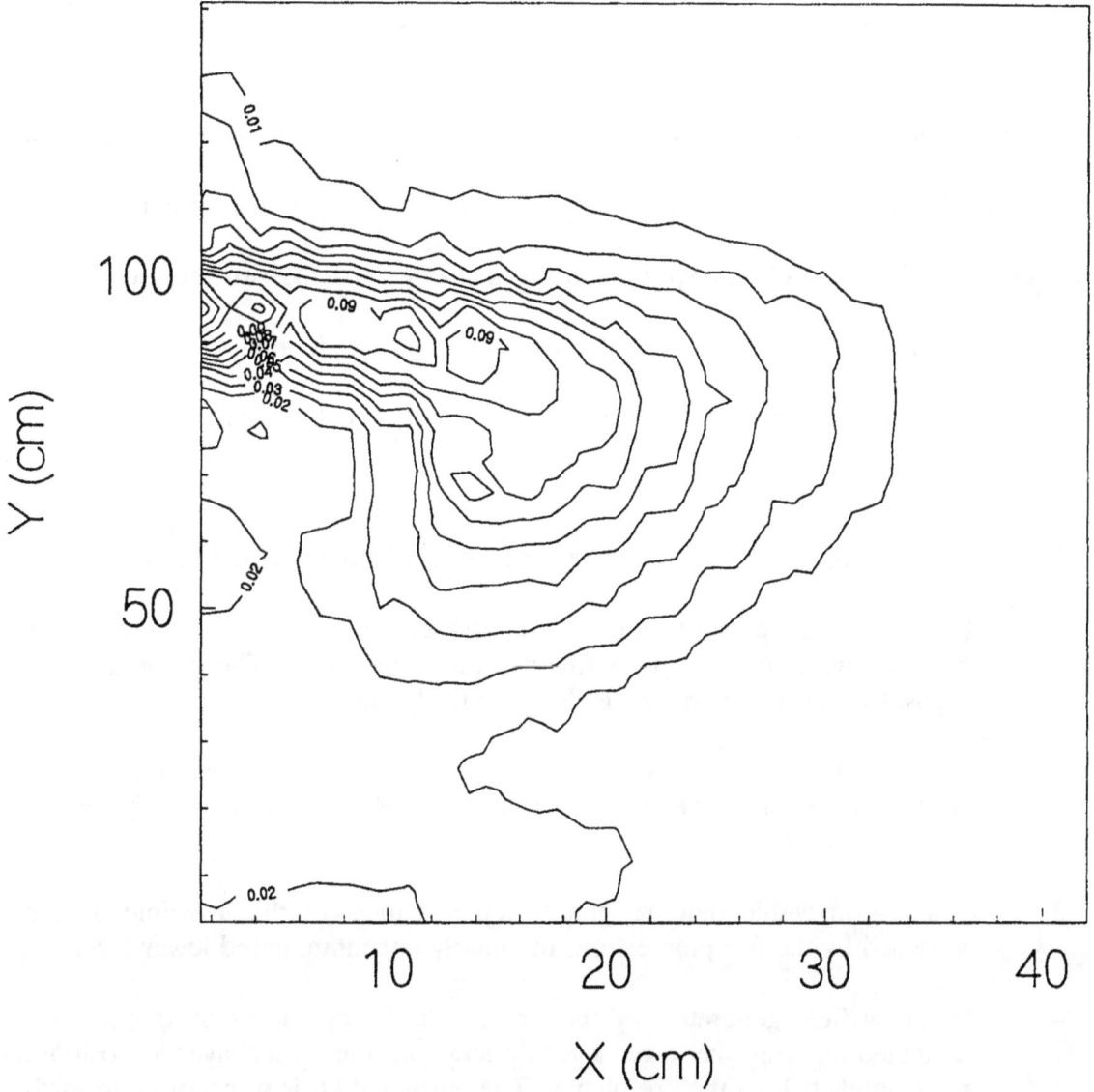

Figure 73.    Oxygen concentration contours calculated by FLOW3D for a horizontal plane located 63 cm above the floor of the RSE are shown for a 400 kW natural gas fire. The axis system has its origin in the center of the front of the enclosure. y is parallel to the long axis of the enclosure and x lies parallel to the short side. Only half of the short side is represented since negative values of x are mirror images. Positive values of x are truncated as the enclosure actually extends from 0 to 49 cm.

the rear of the fire plume as will become obvious in the next paragraph. The flow patterns responsible for the transport of $O_2$ can be seen in figure 71.

Calculated $O_2$ concentrations for a vertical cut along the RSE are shown in figure 74 for a position 1.5 cm from the centerline. In the lower layer the air away from the fire plume is nearly undiluted which agrees with experimental observations. A region of entrainment of air ($O_2$) into the upper layer behind the fire plume can be seen in this plane. It s this $O_2$ which is ultimately transported to the front of the RSE. In the model, this oxygen reacts to form $CO_2$ and $H_2O$. The detailed chemical-kinetic modeling discussed in Section IV suggests that such entrained $O_2$ would actually react in the upper layer to generate CO, $H_2$, and $H_2O$.

The principal findings of the field modeling investigation can be summarized as:

1.      There is a recirculation zone within the upper layer in the rear of the enclosure which consists primarily of combustion gases. The residence time of gases in this region of the RSE is relatively long.

2.      Since combustion gases are exiting the doorway at a high rate, the residence times for upper-layer gases in the front of the RSE are considerably shorter than in the rear of the enclosure.

3.      Combustion within the fire plume depletes most of the available oxygen entrained by the fire plume from the nearly uncontaminated lower layer.

4.      The flow field generated by the fire results in significant entrainment of unvitiated air from the lower layer directly into the upper layer at positions immediately behind the fire plume. The entrained $O_2$ is transported towards the front of the RSE by the fire-induced flow. This behavior is supported by the calculated velocity contours as well as the higher temperatures calculated for regions in the front of the enclosure.

# Oxygen Mole Fraction

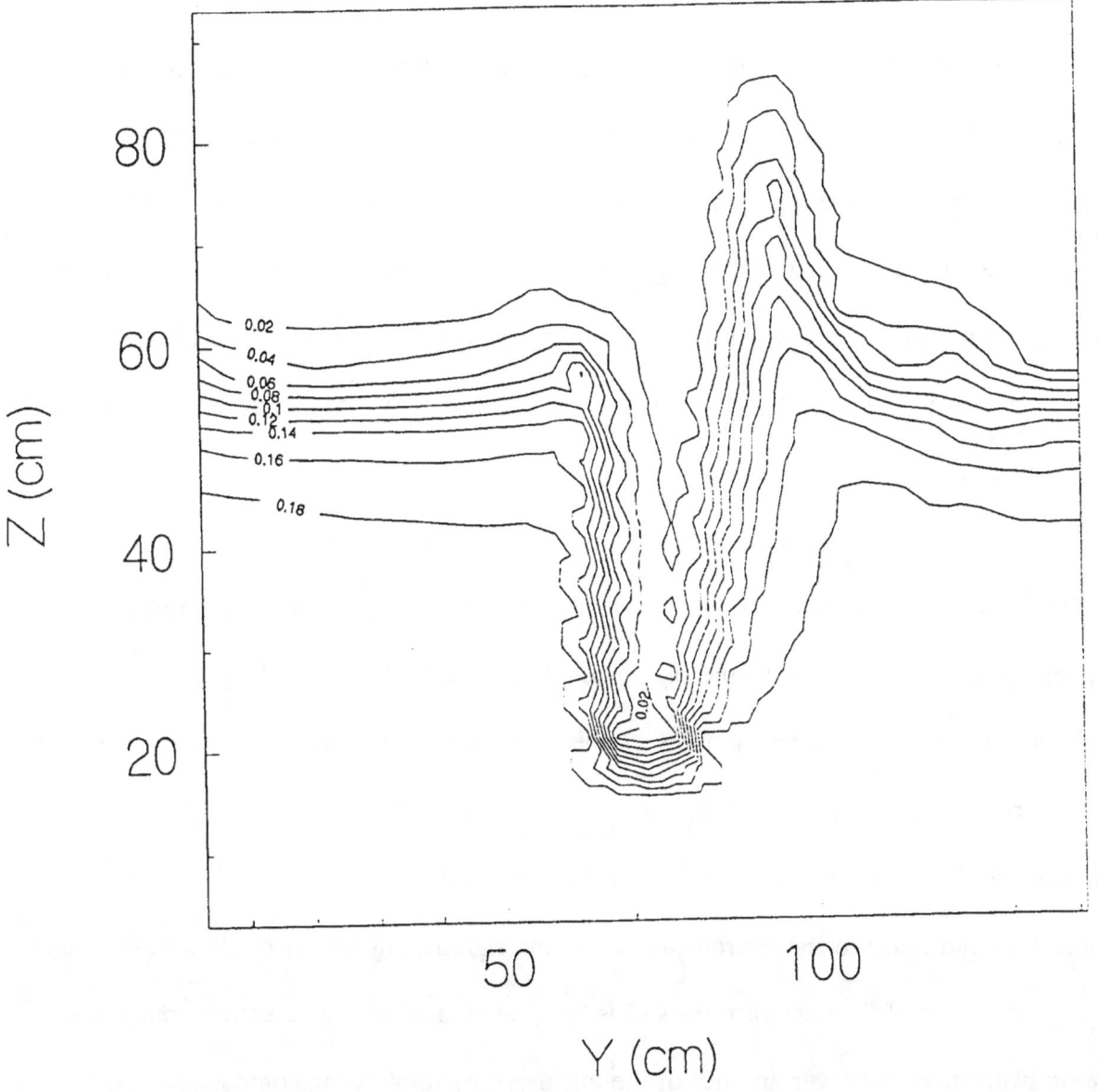

Figure 74. Oxygen concentration contours calculated by FLOW3D for a vertical plane aligned along the long axis and 1.5 cm from the centerline of the RSE are shown for a 400 kW natural gas fire. The axis system origin is at the bottom and 1.5 cm from the center of the doorway located in the front of the enclosure. z is the vertical direction and y is parallel to the long side of the enclosure. The fuel source is located 73 cm from the doorway at the center of the enclosure.

135

5.      Conclusions Based on the Experimental and Modeling Studies of the RSE.


By comparing the modeling and experimental findings for the RSE it is possible to make hypotheses concerning the causes for many of the most interesting observations. Of course, careful experimental work will be required to confirm these hypotheses fully.

The two most relevant findings from the k-$\epsilon$ modeling study are that there is a recirculation zone in the rear of the enclosure for which the residence time of combustion gases is relatively long, and that significant air is mixed directly into the highly vitiated upper layer from the lower layer.

Residence time arguments can be used to explain the observed differences in CO concentration between the rear and front of the enclosure when wood is present in the upper layer. Based simply on the higher temperatures in the front of the enclosure, the generation rate of gases by pyrolysis of the wood should be faster in the front of the enclosure. However, experimentally the CO levels are considerably higher in the rear. Even though the CO is being produced in the rear of the enclosure more slowly, it is spending more time there due to the recirculation zone and significantly higher concentrations build up. In the front of the enclosure the CO is being generated at a greater rate, but it is also being swept from the layer and out of the enclosure by the flowing combustion gases.

The experiments with only natural-gas fuel have shown that CO concentrations as a function of $\phi_g$ observed in the rear of the enclosure are very similar to those measured by Toner [22] in his hood experiments (see figures 56-58). Temperatures in the rear of the enclosure are high enough that comparison with Toner's data is more appropriate than with

136

Morehart's results [23]. The close agreement shows that the combustion gases which are trapped in the rear of the enclosure have generation rates for individual species which are in the same relative proportions as in Toner's experiments. This is strong evidence that the relative generation rates of combustion gases by the fire plume are identical in the enclosure and hood experiments. The same conclusion was reached when the experimental findings for the enclosure fires of Roby and coworkers [42],[43] were compared with the hood experiments of Beyler [16],[17],[18].

The reduced-scale enclosure fires burned at NIST have also shown that there are significant increases in CO concentrations in the upper layer near the ceiling as one moves from the fire source towards the doorway. This observation suggests that reactions are occurring in the upper layer. The k-$\epsilon$ modeling provides an explanation. The modeling shows that air is mixed directly into the upper layer for these positions. Two fates can be imagined for this air when it enters the heated rich gases typical of the upper layer. It could react with rich gases and burn as a normal diffusion flame without rapid mixing and would be expected to produce primarily $H_2O$ and $CO_2$. A second possibility is that the air becomes rapidly mixed with the combustion gases and the $O_2$ concentration is decreased rapidly below the flammability limits for the fuel-air mixture. Under these conditions, reactions similar to those calculated in the detailed chemical-kinetic analysis [29] are expected, and CO should be formed in preference to $CO_2$. The observation of increases in CO concentration as one moves toward the front of the enclosure indicates that the latter possibility is the correct one. This explains why CO concentrations higher than predicted by the GER concept are found.

The RSE experiments have also shown that there are significant changes in CO concentration with height in the front of the upper layer. The field modeling investigation shows that is expected due to calculated variations in the mixing behavior with position in the upper layer. This dependence on position is sufficient to explain the experimental observations.

The conclusion that air enters the front of the upper layer by a direct pathway, in addition to entrainment into the fire plume, suggests that for underventilated burning, values of upper-layer $\phi_\ell$ are lower in the front of the enclosure than in the rear. Since $O_2$ concentrations are near zero in both the front and read, additional chemical reaction must occur for gases reaching the upper layer in the front of the enclosure. These reactions generate heat, thus providing an explanation for the higher upper-layer temperatures observed within the upper layer in the front of the enclosure as compared to the rear.

Since more air is mixed into the upper layer for the front of enclosure, it is expected that values of $\phi_\ell$ will also be lower in the front than in the rear. Evidence for this behavior can be found in figure 52. In the rear of the enclosure the oxygen concentration falls to zero for $\phi_g \approx 1.0$, while for measurements in the front this occurs for $\phi_g \approx 1.6$ indicating that, for a given $\phi_g$, the fuel/air ratio is indeed lower in the front.

The evidence therefore suggests that the generation rate of CO for a buoyancy-driven fire plume entering a vitiated upper layer within an enclosure can be correlated in terms of the upper-layer global equivalence ratio if proper account is taken of the temperature effect. This has been demonstrated for the low and high temperature limits in the RSE experiments. Two additional mechanisms have been identified which are capable of increasing the concentration of CO in the upper layer, and which have not been experimentally characterized previously. The first requires rapid mixing of very lean gases (i.e., air) directly into a rich, high-temperature upper layer where reactions similar to those for a flow reactor take place. Due to the rich nature of the layer, the principal oxidized product of these reactions is CO in preference to $CO_2$. The second mechanism is the special, but common, case of an enclosure fire where wood, or, possibly, other polymeric fuels containing oxygen atoms, is located within the upper layer. When the upper layer is depleted of $O_2$, effectively whenever there is a rich upper layer at temperatures > 900 K, the pyrolysis of wood will lead to significant increases in CO concentration.

## VI. VALIDITY OF USING THE GLOBAL EQUIVALENCE RATIO CONCEPT TO PREDICT CARBON MONOXIDE PRODUCTION IN ENCLOSURE FIRES

The experimental and modeling work summarized above allows an assessment of the conditions for which the use of the GER concept is appropriate for predicting CO formation in enclosure fires, as well as conditions for which the approach is not expected to work. These different regimes are summarized here. They are listed in order of priority. Note that this discussion assumes that wood or other pyrolyzing solids are not present in the upper layer. In these cases, the GER concept cannot be applied if the upper layer is rich.

### A. Conditions Where the GER Concept is Appropriate for Enclosure Fire Predictions of CO Formation

1. Fires for Which the Upper-Layer Temperature is Less than 700 K

The use of the GER concept is most appropriate for enclosure fires falling into this category. The upper layers are expected to be nonreactive and both hood-type and enclosure experiments indicate that the relative production rates of combustion products are very similar for the full range of $\phi_g$.

2. Fires for Which the Upper Layer is Lean and Very Hot (> 900 K)

Fires falling into this category are expected to generate combustion products in amounts very similar to free-burning, fully ventilated fires. In general, CO levels are expected to be low unless the fuel generates unusually high levels of soot.

3. Fires For Which the Only Route for Oxygen (Air) to Enter the Upper Layer is Through the Fire Plume and the Upper Layer is Very Hot (> 900 K)

For this condition the fires will generate combustion products in the manner typical of hood fires having high temperatures over a full range of $\phi_g$. With regard to CO, it has been shown that for $\phi_g > \approx 1.5$, yields are relatively constant and similar to those observed in low-temperature hood experiments.

140

**B.** **Conditions Where the GER Concept is Inappropriate for Enclosure Fire Predictions of CO Formation**

1. Fires Having $\phi_g > 0.5$ and Intermediate Temperatures (700-900 K) in the Upper Layer

Hood experiments have shown that for these conditions the relative generation rates for combustion products in the plume are dependent on the upper-layer temperature. Reactions within the upper layer itself are also possible and will be dependent on the layer residence time. The use of the GER concept for these conditions would be inappropriate.

2. Fires for Temperatures > 900 K for Which Oxygen (Air) Enters a Rich Upper Layer Directly

Both detailed chemical-kinetic analysis and experiments indicate that for these conditions any oxygen which reaches the upper layer directly reacts with the rich gases to generate primarily CO as opposed to $CO_2$. The concentrations of CO observed for these cases is higher than would be predicted based on the GER concept. It is possible that the GER concept can still be used to predict the generation rates of combustion gases by the fire plume, and that an additional model can be employed to predict the changes in concentration as the result of upper-layer reactions.

**C.    Implications for Using the GER Concept to Model CO Formation in Enclosure Fires Where CO Toxicity is Important**

The most important scenario for which CO formation has been implicated in fire deaths is a fully developed, flashed-over enclosure fire where the victims are located in compartments remote from the fire. Presumably, toxic gases (primarily CO) are transported from the fire room to locations where there are people. The discussion above indicates that these fires burn underventilated and have very high upper-layer temperatures. The GER concept can be used to predict combustion gas concentrations in fires of this type only for cases meeting the following criteria:

1.    The fire burns in two well defined layers, and the only pathway for air to enter the upper layer is through the fire plume itself.

2.    There are no fuels capable of being pyrolyzed at high temperatures located in the upper layer. In particular, no wood or other fuel capable of generating CO by pyrolysis is present in the upper layer.

The RSE experiments and modeling summarized above have demonstrated that a significant mass of air can enter the upper layer of an enclosure fire directly, and that much higher concentrations of CO are generated than predicted by the GER concept. The amount of direct mixing between the two layers in an intense enclosure fire has not been the subject of previous investigation, so it is not possible to generalize the findings of the current investigation. However, it seems likely, based on the RSE results, that such mixing will be characteristic of many real fires in buildings. This is particularly true with regard to the intense fires responsible for most smoke inhalation fire deaths.

142

Rooms in many buildings have ventilation pathways (such as heating or air conditioning ducts) which can introduce air directly into a rich upper layer formed by a fire. When mixing is intense in the upper layer it is likely that air introduced in this manner will mix rapidly with the rich gases and generate CO by the mechanisms already discussed.

Wood and other polymeric materials are often found on the ceilings and upper walls of enclosures. At the present time it is impossible to predict the CO and other combustion gases generated in these fires, but it is safe to say that when these materials are present in significant amounts, the GER concept will fail to predict accurately generation rates for combustion products.

In a recent publication Babrauskas et al. have recommended yields to be used for the prediction of CO generation by fully developed, underventilated enclosure fires [3]. These recommended yields are derived primarily from observations in the hood experiments. Based on the findings discussed in here, it must be concluded that their estimates for CO yield represent a lower limit, and that significantly higher yields of CO are possible for enclosure fires.

To summarize, there are limited conditions for enclosure fires for which the GER concept will allow accurate predictions of combustion gas generation. There are also likely to be cases for which the GER concept can be used to calculate generation rates for a fraction of the upper-layer gases, and that additional understanding may allow total generation rates to be predicted. However, it is clear that the GER concept, as originally formulated from the hood-type experimental findings, cannot be directly extended to the types of enclosure fires which are most likely to result in death by smoke inhalation.

Additional understanding of very complicated processes (e.g., entrainment of air directly into the upper layer of an intense fire and high temperature pyrolysis of polymers) will be required before accurate predictions of CO can be made for the most relevant enclosure fires.

## VII.   ADDITIONAL DISCUSSION AND LIMITATIONS OF UNDERSTANDING

The formation of CO in enclosure fires has been long been recognized as a crucial element in fire deaths.  However, systematic investigations of CO formation mechanisms and quantification of its generation rate have only been initiated within the past fifteen years.  The development of the GER concept was based on results of the first serious investigations designed to isolate these mechanisms.  The simplified approach of investigating combustion product formation in the hood experiments was not only appropriate, but crucial, for developing an understanding of CO formation in enclosure fires.  The GER concept has provided the impetus, direction, and framework for later experiments and analyses.  It is disappointing, but in retrospect not surprising, that the hood experiments fail to capture the full complexity of enclosure fire behavior.  By building on the GER concept, additional mechanisms for the generation of CO in enclosure fires have been identified, and the areas where additional understanding is required made clearer.

In this report experimental and theoretical modeling investigations have focused on the generation of CO in enclosure fires.  It is important to point out that while significant progress has been made, our understanding of the complete problem is far from that required to make confident predictions of the CO reaching a location in a building well removed from the fire origin. Some of the serious limitations of the current work and areas of uncertainty are summarized here in the hope that this work will not be used in an inappropriate manner to attempt predictions of CO formation for actual enclosure fires.

All of the recent investigations (both hood-type and enclosure fires) discussed have been made using single isolated fuel sources in reduced-scale conditions. Actual enclosure fires often involve multiple fuel sources of different fuel types which are burning throughout the enclosure. The mixing and flow behaviors (which have been shown to be crucial to the generation of products of incomplete combustion) are likely to very different than observed for a single fire plume. As a result, CO generation rates could also be very different. It remains to be demonstrated whether the limited reduced-scale results can be extrapolated to full-scale enclosures.

There are additional mechanisms for the generation and destruction of CO which have not been considered in the investigations thus far. For instance, it is well known that when soot and $CO_2$ are present together in a high temperature environment, they can react to generate CO. For fuels which generate high soot concentrations, this reaction may create even higher concentrations of CO than would be formed by the mechanisms identified thus far.

A very important question concerns the mechanisms and conditions responsible for quenching of the combustion of rich fuel gases exiting an enclosure fire. The reduced-scale enclosure investigations discussed above have found that levels of CO for distances far from the fire are quite low as the result of additional burning outside of the enclosure

145

[42],[43],[44]. In these experiments, the combustion gases exit into open areas where they mix with air and burn intensely. Gottuk et al. have reported a study of this type of external burning on the reduction in CO concentration [62]. Measurements in the furniture calorimeter have demonstrated that the same is true in the RSE experiments [44]. In building fires the combustion gases often exit into adjoining rooms or corridors. Fire experience and full-scale fire tests have shown that high concentrations of CO can be transported to remote locations indicating that quenching of additional combustion outside of the fire room occurred. The quenching process must be characterized before predictions of CO for remote locations can be made. Studies of this type are just being initiated.

## VIII. SUMMARY AND FINAL REMARKS

This paper has summarized and analyzed the findings of a number of investigations designed to characterize the formation of CO and other products of incomplete combustion in enclosure fires. The focus has been on whether or not the GER concept can be used to predict the generation rates for these gases.

It has been found that high levels of CO are formed in enclosure fires which are underventilated. One mechanism responsible for the formation of CO that has been identified and characterized in hood-type experiments is the quenching of a fire plume upon entering an upper layer of rich combustion products. The combustion products generated by this process have been shown to be strongly correlated with the upper-layer equivalence

146

ratio, $\phi_g$. Experiments indicate that the correlations are dependent on the upper-layer temperature, but that well defined correlations exist for low ($< \approx 700$ K) and high ($> \approx 900$ K) temperatures. The findings suggest that shifts in combustion gas compositions are primarily the result of changes in the reaction behavior within the fire plume, despite the fact that the upper layer can also become reactive at the higher temperatures.

Experiments have shown that the GER concept can be extended to enclosure fires for conditions where the fire burns in a similar manner, or when temperatures are low enough ($< 700$ K) to ensure the upper layer is not reactive. In actual enclosure fires, temperatures are generally such that the upper layer can be reactive. In this case, the extension of the GER concept is appropriate only when there are two well-defined layers within the enclosure, no pathways exist for the introduction of air directly into the upper layer, and there are no solids capable of pyrolysis located in the upper layer. Analysis indicates that the conditions necessary for the application of the GER concept are not typical of the enclosure fires most likely to result in smoke inhalation deaths.

Mechanisms for the formation of CO in fires involving direct entrainment of air into a layer of rich combustion gases and the pyrolysis of wood have been demonstrated. The generation of CO by the former mechanism is consistent with predictions of a detailed chemical-kinetic model which indicates the formation of CO will be favored over $CO_2$.

Limitations of the current understanding of CO formation in and transport from an enclosure fire have been briefly discussed.

## IX. ACKNOWLEDGMENTS

This work is the result of efforts of a large number of researchers who have provided material help as well as ideas. Discussions with Nelson Bryner, Bill Davis, Rik Johnsson, and Tom Ohlemiller of NIST were most helpful. Nelson Bryner carefully read the manuscript and provided many comments which substantially improved the final version. Bill Davis and Rik Johnsson were most helpful in providing figures and technical details. Dan Gottuk provided many figures which are included here as well as helpful discussions concerning the VPISU research effort.

## X. REFERENCES

[1]   Harland, W. A.; Anderson, R. A. Causes for death in fires. Proceedings, smoke and toxic gases from burning plastics; pp. 15/1-15/19; Jan. 6-7, 1982, London, England.

[2]   Harwood, B.; Hall, J. R. What kills in fires: smoke inhalation or burns? Fire J. 83(3): 29-34; 1989 May/June.

[3]   Babrauskas, V.; Levin, B. C.; Gann, R. G.; Paabo, M.; Harris, Jr., R. H.; Peacock, R. D.; Yusa, S. Toxic potency measurement for fire hazard analysis. National Institute of Standards and Technology Special Publication 827; 1991 December. 107 p.

[4]   Pitts, W. M. Long-range plan for a research project on carbon monoxide production and prediction. National Institute of Standards and Technology NISTIR 89-4185; 1989 October. 40 p.

[5]   Pitts, W. M. Executive summary for the workshop on developing a predictive capability for CO formation in fires. National Institute of Standards and Technology NISTIR 89-4094; 1989 May. 68 p.

[6]   Mulholland, G.; Janssens, M.; Yusa, S.; Twilley, W.; Babrauskas, V. The effect of oxygen concentration on CO and smoke produced by flames. Cox, G.; Langford, B. eds. Fire safety science--proceedings of the third international symposium. New York NY: Elsevier; 1991. 585-594.

[7] Morehart, J. H.; Zukoski, E. E.; Kubota, T. Characteristics of large diffusion flames burning in a vitiated atmosphere. Cox, G.; Langford, B. eds. in Fire safety science-- proceedings of the third international symposium. New York, NY: Elsevier; 1991. 575-583.

[8] Morehart, J. H.; Zukoski, E. E.; Kubota, T. Chemical species produced in fires near the limit of flammability. Fire Safety J. 19(2 and 3): 177-188; 1992.

[9] Peacock, R. D.; Bukowski, R. W. A prototype methodology for fire hazard analysis. Fire Technol. 26(1): 15-40; 1990 February.

[10] Babrauskas, V. Upholstered furniture room fires--measurements, comparison with furniture calorimeter data, and flashover predictions. J. Fire Sciences 2(1): 5-19; 1984 January/February.

[11] Smith, R. P. Toxic response of the blood, chapter 8 in Casarett and Doull's toxicology--the basic science of poison. Klassen, C. C.; Amdur, M. O.; Doull, J. eds. New York, NY: Macmillan; 1986. 223-244.

[12] Karter, Jr., M. J. Fire losses in the United States during 1991. NFPA J. 86(5): 32-43; 1992 September/October.

[13] Mulholland, G. W. Position paper regarding CO yield. Letter report to Richard G. Gann, Chief, Fire Measurement and Research Division, Center for Fire Research; 1988 June 16. 8p, unpublished.

[14] Levine, R. S.; Nelson, H. E. Full scale simulation of a fatal fire and comparison of results with two multiroom models. National Institute of Standards and Technology Internal Report NISTIR 90-4268; 1990 August. 101 p.

[15] Mulholland, G. W. Comparison of predicted CO yield with results from fire reconstruction of Sharon, PA fire. Letter report to William Pitts, Leader, CO Priority Project; 1989 November 1. 13 p.

[16] Beyler, C. L. Development and burning of a layer of products of incomplete combustion generated by a buoyant diffusion flame. Doctor of Philosophy Thesis, Harvard University; 1983 September. 162 p.

[17] Beyler, C. L. Major species production by diffusion flames in a two-layer compartment fire environment. Fire Safety J. 10(1): 47-56; 1986 January.

[18] Beyler, C. L. Major species production by solid fuels in a two layer compartment fire environment. Grant, C. E.; Pagni, P. J. eds. Fire safety science--proceedings of the first international symposium. New York, NY: Hemisphere; 1991. 431-440.

[19]    Beyler, C. L. Ignition and burning of a layer of incomplete combustion products. Combust. Sci. Tech. 39: 287-303; 1984.

[20]    Cetegen, B. M. Entrainment and flame geometry of fire plumes. Doctor of Philosophy Thesis, California Institute of Technology. 1982 May. Also available as: Cetegen, B. M.; Zukoski, E. E.; Kubota, T. Entrainment and flame geometry of fire plumes. National Bureau of Standards Government Contractor's Report GCR-82-402; 1982 August. 184 p.

[21]    Lim, C. Entrainment in fire plumes. Part 2. Masters Thesis. California Institute of Technology; 1984. See also: Zukoski, E. E.; Kubota, T.; Lim, C. S. Experimental study of environment and heat transfer in a room fire. Mixing in doorway flows and entrainment in fire plumes. National Bureau of Standards Government Contractor's Report GCR-85-493; 1985 May. 119 p.

[22]    Toner, S. J. Entrainment, chemistry and structure of fire plumes, Doctor of Philosophy Thesis, California Institute of Technology; 1986. 259 p. Also available as: Toner, S. J.; Zukoski, E. E.; Kubota, T. Entrainment, chemistry and structure of fire plumes. National Bureau of Standards Government Contractor's Report GCR-87-528; 1987 April. 222 p.

[23]    Morehart, J. H. Species produced in fires burning in two-layered and homogeneous vitiated environments. Doctor of Philosophy Thesis, California Institute of Technology; 1990. Also available as: Morehart, J. H.; Zukoski, E. E.; Kubota, T. Species produced in fires burning in two-layered and homogeneous vitiated environments. National Institute of Standards and Technology Government Contractor's Report GCR-90-585; 1990 December. 259 p.

[24]    Zukoski, E. E.; Toner, S. J.; Morehart, J. H.; Kubota, T. Combustion processes in two-layered configurations. Grant, C. E.; Pagni, P. J. eds. Fire safety science-- proceedings of the first international symposium. New York, NY: Hemisphere; 1988. 295-304.

[25]    Zukoski, E. E.; Morehart, J. H.; Kubota, T.; Toner, S. J. Species production and heat release rates in two-layered natural gas fires. Combust. Flame 83(3 and 4): 325-332; 1991 February.

[26]    Cooper, L. Y. A model for predicting the generation rate and distribution of products of combustion in two-layer fire environments. National Institute of Standards and Technology Internal Report NISTIR 4403; 1990 September. 47 p.

[27]    Cooper, L. Y. Applications of the generalized global equivalence ratio model (GGERM) for predicting the generation rate and distribution of products of combustion in two-layer fire environments--methane and hexanes. National Institute of Standards and Technology Internal Report NISTIR 4590; 1991 June. 69 p.

[28]     Cooper, L. Y. A model for predicting the generation rate and distribution of products of combustion in two-layer fire environments.   Yao, S. C.; Chung, J. N., ed. Proceedings of the meeting on heat and mass transfer in fires and combustion systems--1991, HTD-vol.176; 1991, December 1-6; New York, NY: The American Society of Mechanical Engineers; 1991. 9-22.

[29]     Pitts, W. M. Reactivity of product gases generated in idealized enclosure fire environments.  Twenty-fourth symposium (international) on combustion; Pittsburgh, PA: The Combustion Institute; 1992. 1737-1746.

[30]     Kee, R.J.; Miller, J.A.; Jefferson, T.H. CHEMKIN: a general purpose, problem-independent, transportable, Fortran chemical kinetics code package. Sandia National Laboratories Report SAND80-8003; 1980 March.  197 p.

[31]     Lutz, A. E.; Kee, R. J.; Miller, J. A. SENKIN: A Fortran program for predicting homogeneous gas phase chemical kinetics with sensitivity analysis. Sandia National Laboratories Report SAND87-8248; 1988 February. 27 p.

[32]     Glarborg, P.; Kee, R. J.; Grcar, J. F.; Miller, J. A. PSR: a Fortran program for modeling well-stirred reactors. Sandia National Laboratories Report SAND86-8209; 1986 February. 46 p.

[33]     Dagaut, P.; Boettner, J. -C.; Cathonnet, M.  Int. J. Chem. Kin. 22(6): 641-664; 1990 June.

[34]     Kee, R. J.; Rupley, F. M.; Miller, J. A. The Chemkin thermodynamic data base. Sandia National Laboratories Report SAND87-8215; 1987 April. 81 p.

[35]     Nakaya, I. Prediction model of CO, $CO_2$ and $O_2$ concentrations in compartment fires using wood fuel.  Fire Materials 173(4): 173-178; 1987 December.

[36]     Rasbash, D. J.; Stark, B. W. V. The generation of carbon monoxide by fires in compartments.  Fire Research Station F.R. Note No. 614; 1966 February. 38 p.

[37]     Gross, D.; Robertson, A. F. Experimental fires in enclosures. National Bureau of Standards Report 8147; 1963 December 12. 24 p.

[38]     Gross, D.; Robertson, A. F. Experimental fires in enclosures. Tenth symposium (international) on combustion. Pittsburgh, PA: The Combustion Institute; 1965. 931-942.

[39]     Tewarson, A. Some observation on experimental fires in enclosures. Part I: cellulosic materials. Combust. Flame 19(1): 101-111; 1972 August.

[40]    Tewarson, A. Fully developed enclosure fires of wood cribs. Twentieth symposium (international) on combustion; Pittsburgh, PA: The Combustion Institute; 1984. 1555-1566.

[41]    Morikawa, T.; Yanai, E. Toxic gases evolution from air-controlled fires in a semi-full scale room. J. Fire Sciences 4(5): 299-314. 1986 September/October.

[42]    Gottuk, D. T. The generation of carbon monoxide in compartment fires. Doctor of Philosophy Thesis, Virginia Polytechnic Institute and State University; 1992 September. Also available as: Gottuk, D. T. Generation of carbon monoxide in compartment fires. National Institute of Standards and Technology Government Contractor's Report NIST-GCR-92-619, 1992 December. 265 p.

[43]    Gottuk, D. T.; Roby, R. J.; Peatross, M.; Beyler, C. L. Carbon monoxide production in compartment fires. J. Fire Protection Eng. 4(4): 133-150. 1992 October-December.

[44]    Bryner, N.; Johnsson, R. ; Pitts, W. M. Carbon monoxide production in compartment fires--reduced-scale enclosure test facility. To appear as a National Institute of Standards and Technology Internal Report.

[45]    Proposed standard method for room fire test of wall and ceiling material assemblies, Annual Book of ASTM Standards, Part 18, American Society for Testing and Materials; Philadelphia, PA. 1982.

[46]    Fire tests--full scale room test for surface products. Draft International Standard ISO/DIS 9705; International Organization for Standardization. 1990.

[47]    Kawagoe, K. Fire behavior in room. Report of the Building Research (Japan). 1958 September. 72 p.

[48]    Heskestad, G. Modeling of enclosure fires. Fourteenth symposium (international) on combustion, Pittsburgh, PA: The Combustion Institute; 1972. 1021-1030.

[49]    Quintiere, J. G. Scaling applications in fire research. Fire Safety J. 15(1): 3-29; 1989.

[50]    Babrauskas, V. Upholstered furniture heat release rates: measurements and estimation. J. Fire Sciences 1(1): 9-32; 1983 January/February.

[51]    Steckler, K. D.; Baum, H. R.; Quintiere, J. G. Fire induced flows through room openings--flow coefficients. Twentieth symposium (international) on combustion; Pittsburgh, PA: The Combustion Institute; 1984. 1591-1600.

[52]    Janssens, M.; Tran, H. C. Data reduction of room tests for zone model validation. J. Fire Sciences 10(6): 528-555. 1992 November/December.

[53]    Babrauskas, V.; Parker, W. J.; Mulholland, G. W.; Twilley, W. H. The phi-meter: a simple, fuel-independent instrument for monitoring combustion equivalence ratio. Accepted for publication in Review of Scientific Instruments.

[54]    Hileman, F. D.; Wojcik, L. H.; Futrell, J. H.; Einhorn, I. N. Comparison of the thermal degradation products of $\alpha$-cellulose and Douglas fir under inert and oxidative environments. in Thermal uses and properties of carbohydrates and lignins. Shafizadeh, F.; Sarkanen, K. V.; Tillman, D. A., eds. New York, NY: Academic Press; 1976. 49-71.

[55]    Arpiainen, V.; Lappi, M. Products from the flash pyrolysis of peat and pine bark. J. Anal. Appl. Pyrolysis 16(4): 355-376; 1989.

[56]    Pitts, W. M.; Bryner, N. P.; Johnsson, E. L. Carbon monoxide formation in fires by high-temperature anaerobic wood pyrolysis. Twenty-fifth symposium (international) on combustion. accepted for publication.

[57]    Davis, W. D. Analysis of a reduced scale enclosure using a field model. National Institute of Standards and Technology Internal Report; 1993, to appear.

[58]    CFD Department, AEA Industrial Technology, Harwell Laboratory, Oxfordshire, U. K. Harwell-FLOW3D Release 2.3: Users Manual, 1990 July.

[59]    Davis, W. D.; Forney, G. P.; Klote, J. H. Field modeling of room fires. National Institute of Standards and Technology Internal Report 4673; 1991 November. 40 p.

[60]    Kumar, S.; Gupta, A. K.; Cox, G. Effects of thermal radiation on the fluid dynamics of compartment fires. Cox, G.; Langford, B. eds. Fire safety science--proceedings of the third international symposium. New York NY: Elsevier; 1991. 345-354.

[61]    Spalding, D. B. Development of the eddy-break-up model of turbulent combustion. Sixteenth symposium (international) on combustion. Pittsburgh, PA: The Combustion Institute; 1977. 1657-1663.

[62]    Gottuck, D. T.; Roby, R. J.; Beyler, C. L. A study of carbon monoxide and smoke yields from compartment fires with external burning. Twenty-fourth symposium (international) on combustion. Pittsburgh, PA: The Combustion Institute; 1992. 1729-1735.

## XI.  EXECUTIVE SUMMARY

This paper summarizes the findings of a number of experimental and computational

investigations which deal with the formation of carbon monoxide (CO) in fires.  These

summaries and a critical analysis of the findings are used to answer the question:

> Can the generation behavior of CO observed in hood experiments designed
> to model two-layer burning be extended to predict CO generation in actual
> enclosure fires?

The work is justified in terms of the pivotal role that CO formation and transport during

enclosure fires assumes in fire deaths.

A group of experiments which have investigated the formation of CO in well-defined

two-layer environments, designed to model the similar environment often associated with

enclosure fires, are discussed.  In these experiments the products of combustion for fires

burning in open laboratories were trapped in hoods where the concentrations of the gases

were measured.  These "hood" experiments showed that the formation of major products of

combustion can be correlated with the global equivalence ratio (GER), which is defined as

the mass ratio of gases in the hood derived from fuel and air divided by the ratio required

for stoichiometric burning.  The existence of these correlations is referred to as the *GER*

*concept*.  A dependence of the correlations on gas temperature in the hoods is discussed.

An investigation of the reactivity of the upper layer gases using full-kinetic modeling

is summarized.  The implications of the results for applying the GER concept to enclosure

fires are provided.

Results from a large number of experiments for fires in enclosures which shed light

on the formation of CO are reviewed.  These include large-scale fire tests as well as

154

measurements in reduced-scale enclosures. Particular attention is paid to recent reduced-scale experiments at Virginia Polytechnic University (VPISU) and the National Institute of Standards and Technology (NIST) which were designed investigate the formation of CO by carefully controlled enclosure fires and to allow comparisons with the results of hood experiments. The NIST study included a field modeling calculation designed to characterize the flow patterns within the enclosure.

The VPISU investigation showed that, for conditions where the only pathway for air to enter the combustion gases is through the fire plume, unique correlations of the combustion products with the GER exist, and that these correlations are very similar to those found in hood experiments using the same fuels. The NIST experiments indirectly verified this conclusion. The NIST study also showed, however, that the flow fields typical of enclosure fires are capable of mixing air directly into the upper layer. Such mixing for high upper-layer temperatures leads to the generation of CO as predicted by the full-kinetic modeling. As a result of this effect, the concentrations of CO observed in the upper layer of the fire enclosure are higher than predicted by the GER concept.

Experiments were also done at NIST during which the upper section of walls and the ceiling of the reduced-scale enclosure were lined with wood. A natural gas fire was then burned within the enclosure in a manner identical to the earlier experiments. As a result of the pyrolysis of the wood by the high-temperature upper layer within the enclosure, extremely high levels of CO were measured. This experiment demonstrates for the first time that the pyrolysis of oxygenated polymers in an oxygen-depleted upper layer of an enclosure fire can lead to the formation of significant levels of CO.

The findings of the investigations to date are used to generate criteria for conditions where the GER concept can be used to predict CO generation during enclosure fires and for which its use is expected to be inappropriate. These conditions are:

Use of GER Appropriate:

1) Fires for which the upper-layer temperature is less than 700 K.

2) Fires for which the upper layer is lean and very hot (> 900 K).

3) Fires for which the only route for oxygen (air) to enter the upper layer is through the fire plume and the upper layer is very hot (> 900 K).

Use of GER Inappropriate:

1) Fires having slightly lean or rich upper layers and intermediate temperatures (700-900 K)

2) Fires for temperatures greater than 900 K for which oxygen (air) enters a rich upper layer directly.

3) Fires which generate rich, high-temperature upper layers and which have solid fuels capable of pyrolysis located in the upper layer.

Analysis of the types of enclosure fires which are responsible for the majority of smoke- inhalation deaths suggests that these fires usually burn underventilated, are quite intense, and have achieved flash over. Wood and/or other polymeric materials are often located in the upper areas of the enclosures. As a result, for the types of fires most likely to kill people, it is concluded that the GER concept alone will not be adequate to predict the production of CO.

● U.S. GOVERNMENT PRINTING OFFICE: 1994-300-633/12686